Dwight E. Gray

So You Have to Write a Technical Report

Elements of Technical Report Writing

I R P Information Resources Press
WASHINGTON, D.C.   1970

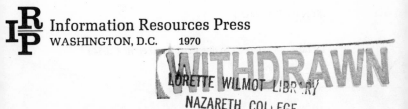

Available from
Information Resources Press
2100 M Street, N.W.
Washington, D.C. 20037

ISBN 0-87815-002-1

Library of Congress Catalog Card Number 70-120541

167901

Preface

Do you find yourself in the following situation?

1. You are under pressure from your boss—or your conscience—to write a technical report.
2. You find the prospect somewhat distasteful, and maybe even a bit frightening.
3. You would welcome suggestions and counsel on getting the project under way.

If so, you may find this book helpful. It is hoped that its several chapters, first, will demonstrate that your report-writing task is less formidable than you may have thought, and, second, will provide you with basic guidance in:

1. Identifying the appropriate audience for your report;
2. Organizing your material and planning the document; and
3. Starting the writing smoothly, carrying it out effectively, and terminating it gracefully.

The term "basic guidance," used above, accurately describes the function envisioned for this volume. It is not a technical-writing "cookbook" of hard-and-fast rules that will relieve you of the necessity of thinking. Some of you may find this lack of precise, rigorous instructions disturbing. But report writing is not like fudge making. No two technical report situations are identical and there is no standard "recipe" that can be automatically applied in all cases. There are, however, valid principles applicable to the preparation of all technical reports; it is with these this book is primarily concerned.

One other thing this book is *not* should be mentioned. It is not a rule book on grammar, sentence structure, style, syntax, accepted usage, and the like, although these topics do receive some attention in the general text. Many excellent treatises on these subjects are already available, representative lists of which are given in Appendixes A and B.

In short, this book is meant to provide you with a kind of "travel guide" to effective technical report writing. Main routes are identified and discussed, and their advantages and limitations are pointed out. Certain good alternate trails are also described, and the conditions are noted under which they are passable and on occasion even preferable. Let us emphasize again, however, that this book will not, nor is it intended to, relieve you of the necessity of thinking while planning and writing your technical report. The late J. Robert Oppenheimer used to counsel his staff at the Institute for Advanced Study, "Never try to speak more clearly than you think." The admonition is equally sound if "write" is substituted for "speak."

Acknowledgments

This volume is based primarily on lectures I presented in technical report writing workshops held for several summers at The Pennsylvania State University. I am particularly grateful, therefore, for the encouragement I received at that time from Dr. Eric A. Walker, president of the University, who proposed the workshops, and for the advice and assistance of my two colleagues in these courses, Christian K. Arnold and Arthur G. Norris. In connection with preparation of the final manuscript for the book, I am deeply indebted to Richard H. Belknap for his suggestions and for his permission to use certain material from the National Academy of Sciences' *Guide for Preparing Manuscripts*; to Constance Carter of the Library of Congress for her substantial help with the bibliographies on style and usage, and on technical writing; and to Betty Cacciapaglia of Information Resources Press for the excellent job of copy editing that reduced to a minimum the instances in which I violated my own rules.

D. G.

Contents

CHAPTER **1**

Nature and Anatomy of a Technical Report

This chapter in effect sets the scene for what follows. It presents:

First—A defining description of technical report as the term is used in this book

Second—A list of the principal parts of a representative technical report

Third—A synopsis of the plan of presentation for the rest of the volume

WHAT IS A TECHNICAL REPORT?

The term technical report identifies it as belonging to the large and heterogeneous family of scientific and technical literature. But what distinguishes it from other members of this clan? Discussed below are four factors that show how the technical report has an identity all its own: principal function of a technical report; characteristics implied by this function; reader groups for whom this function is performed; and the several forms the technical report can take.

Principal Function of a Technical Report

Different kinds of writing have different missions. Novels are often written to entertain, sometimes to inform, occasionally to influence. Political writing, advertising, and propaganda in general are intended to influence the reader to think something, do something, or both. On occasion such writing is entertaining, even if not meant to be; it may also be informative, although seldom either completely or objectively so. Inspirational writing is also directed toward influencing readers' beliefs and actions.

The primary mission of all scientific and technical writing is to inform, although in particular cases such secondary aims as entertaining and influencing may also be important. Perhaps the feature that most clearly distinguishes the technical report from other forms of technical writing is the extent to which the objective of being informative outweighs all others. All good technical reports present data and results; many contain authors' conclusions; frequently, recommendations are also included. Although the last of these is obviously directed toward influencing someone's future action, the principal, and by far the predominant, function of a technical report is simply to inform.

Characteristics Implied by This Principal Function[1]

First, a technical report must be based on factual data if it is to be a member in good standing of this branch of the scientific writing family. It is not a legitimate medium for the author to employ for expressing unsupported personal opinions, promoting pet crusades, or slaying real or imaginary dragons.

Second, effective performance of the technical report's principal mission makes it imperative that the writing be clear, concise, precise, and phrased in unambiguous, functional English. The author should recognize and exploit the inherent power and glory of the simple declarative sentence. He should be especially careful to shun the long, meandering sentence that is crammed with obscurely related phrases and clauses, in which the subject and predicate seem hopelessly to be searching for each other and the verb has a kind of unidentified floating object. Ideally, the report should contain everything necessary for the reader to get the message of the document—and nothing more.

Third, it is important that the pattern of organization of a technical report be logical, have continuity, promote the document's primary mission of conveying information, and be relatively easy for the reader to follow. No single scheme is optimum for all technical reports. In any given case, the author should select the one that he believes will best satisfy all of the above criteria.

---

[1] These characteristics are discussed in greater detail in Chapter 9.

Technical Report Readership

Two particularly significant features mark the relationship between the technical report and its "consumer public." First, the reader can be identified. Typical reader groups within the author's own organization include his immediate supervisor, his professional colleagues, the director of research, and the president and managing board. Outside his organization, his audience may be a contracting agency, other scientists or engineers in the same field, or the general technical community. When, as frequently happens, a single report has to meet the needs of several audiences, the writing problem becomes more complicated. Nevertheless, it is still possible to identify with fair precision at least the principal readership toward whom the report is to be directed.

The other important defining relationship of technical report readers to technical reports is their direct need for the information in the reports written for their use; very often, in fact, they have a right to demand it. For example, the author's supervisor, the director of research, the company president, and the contracting agency certainly have both the need for, and the right to demand, the information.

Types of Technical Report

Technical reports can be separated into categories on several different bases. One common pattern depends upon whether the document is a complete unit in itself (a research report); treats only one phase of a continuing investigation (a progress report); or, at the end of a project, summarizes a series of progress reports (a final report).

Another pattern, based on the degree of formality and elaborateness of presentation, is one that divides these documents into long (or formal) reports, short (or memorandum) reports, and letter reports. Typically, a document of the first kind is fairly lengthy, covers a sizable body of material, and follows a formal scheme of organization. The memorandum report, on the average, is shorter, less comprehensive, and less formally organized. The letter report usually concerns a still smaller segment of information, which it treats even less comprehensively and often very informally.

There is, of course, no inherent correlation between the respective groups in the two breakdowns. At least theoretically, any category of either scheme can include reports of all three types defined in the other. This book is concerned primarily and directly with research reports of the long (or formal) type. However, almost all of the fundamental principles that relate to the preparation of these documents apply to the less formal categories as well. The reader will have no difficulty modifying suggested procedures for the less elaborate kinds of technical report.

PRINCIPAL PARTS OF A TECHNICAL REPORT

Before starting to build a bridge, construct a barbecue pit, or assemble Junior's jungle gym, one should have a picture of the finished product clearly in mind. This section is intended to fulfill for the technical report author the function performed in the above cases by, respectively, the engineering drawing, a photograph torn from a home-and-garden magazine, and the sketch in the assembly manual which invariably is located at the bottom of the packing box. In short, the principal sections of a representative technical research report of the long kind are listed below in the sequence in which the reader might expect to encounter them in the printed document. These sections are:

1. Title and Title Page
2. Summary of Conclusions and Recommendations
3. Table of Contents
4. List of Tables and/or List of Illustrations
5. Abstract
6. Introduction
7. Body of Report (not itself a head; includes procedures, data, results, and the like, with the actual heads varying with the particular report)
8. Conclusions
9. Recommendations
10. Appendixes
11. Bibliography or List of References

It is important that three facts be emphasized at the close of this sketch of the anatomy of a technical report. They are:

1. The pattern presented above is *representative*, not hard and fast, for a research report of the long kind.
2. Not all long reports should necessarily contain

all of these parts; very few reports, however, will need any others.

3. The sequence shown above indicates the order in which the several sections might appear in the finished report, not the order in which they should be written.

PLAN OF PRESENTATION FOR CHAPTERS THAT FOLLOW

This chapter has dealt with the nature and anatomy of the technical report in its completed form. The next chapter is devoted to a series of important steps the author of such a document should take before he begins to put words and sentences on paper. Subsequent chapters consider in detail the preparation of the various report sections—*this time in the order in which we recommend that they usually be written.*

CHAPTER 2

Getting Ready to Write
a Technical Report

Your approach to the task of writing a technical report is likely to take you through the following stages:

First—You recognize, perhaps a little reluctantly, that the report must be written.

Second—You put off starting the task just as long as you can.

Third—You finally decide you can procrastinate no longer, and steel yourself for the job ahead.

Fourth—You plunge into a frenzy of committing words, sentences, and paragraphs to paper.

The tendency to think that nothing is being accomplished until pen is put to paper, or fingers to typewriter, is natural, and requires great willpower to resist. But resist it you must if you hope to produce a first-class report. Extreme haste at this point is no virtue. Unless the writing is based on careful preliminary planning, the completed manuscript is likely to be illogically organized, beset with discontinuities in thought, badly written, and generally ineffective. Several significant elements of the total task fall between the third and fourth stages of the above sequence. Specifically, in getting ready to write you should:

1. Analyze the overall problem.
2. Assemble the data you wish to present.
3. Prepare a working outline.
4. Sit and think.

The rest of this chapter is devoted to detailed consideration of these four pre-writing exercises. Although they are listed and discussed here as separate, discrete acts, when you are ready to perform them you will find that they tend to overlap and merge into each other.

ANALYZING THE OVERALL PROBLEM

Analyzing the report problem that faces you consists in asking yourself, and thoughtfully answering, a number of questions about the impending task. Four queries you should always pose to yourself are:

1. Exactly for whom am I writing this report?
2. Just what information do I wish to convey to this reader or group of readers?

3. How much background dare I assume is possessed by the reader or readers?

4. At what technical level shall I peg the report?

Several possible reader groups were mentioned in Chapter 1; there are, of course, others. The accompanying table suggests some report characteristics related to the above questions for three classes of readers—the report author's supervisor, an upper echelon administrative official, and his fellow engineers or scientists. For each report you must decide precisely who constitutes its primary reading public— the principal groups you wish to tell something. Until you do this, you are in no position to make an intelligent selection of the material and to organize it effectively.

You may find it helpful to integrate this specialized public in your own mind into a single READER who can be visualized as a specific person for whom you are writing. This tactic is analogous to the frequent practice by successful public speakers of picking out two or three members of the audience with whom they then imagine they are conversing directly. If your report has to meet the needs, or demands, of several reader groups with differing major interests, you may find you have to assign your personalized READER a somewhat odd combination of characteristics. It is in just such a situation, however, that it is particularly important for you to recognize these varying reader specifications and to plan your report accordingly.

Reaching a sound decision on this first question is, of course, a fundamental prerequisite to answering the others satisfactorily. To give your reader far more information than is necessary for his, and your, purposes is to bury what he needs in a mass of irrelevant or semi-relevant material and to obscure the emphasis you wish to achieve. To give him insufficient information or the wrong information for his needs is worse. A similar situation prevails regarding the extent of reader background you can assume. Ideally, as noted earlier, your reader should be given all the underlying information required for him to get the message, and no more. But the boundary between "more than necessary" and "not quite enough" can seldom be drawn precisely. It probably is preferable to err a little on the side of too much. Optimum technical level, like necessary background, varies widely for different cases, as is evident even from the three report situations covered in the table. This factor must also be considered very carefully if your report is to have maximum effectiveness.

ASSEMBLING THE DATA

After you have decided which kinds and items of information—new and background—you will include in your report, you should assemble these materials. This step may appear too obvious to require mentioning. The need for performing it systematically does warrant emphasis, however. This stage of the report-writing process is analogous to that performed by the stereo buff who assembles all of the components before starting to build a new amplifier.

Sources of the raw material for your report will

COMPARISON OF REPORT CHARACTERISTICS FOR SEVERAL READER GROUPS

Reader Class	Typical Purposes	Probable Extent of Reader Background	Technical Level at Which Report Should be Written
From: Engineer or scientist *To:* Immediate supervisor	1. To keep supervisor informed of progress; may be issued as a series. 2. To report in detail on a particular project—often for supervisor's use in preparing a broader report. 3. To recommend initiating work on a new project, or on a	1. Relatively complete for purposes Nos. 1 and 2. 2. Less complete for purpose No. 3, particularly if a highly specialized new effort is being recommended.	1. Usually relatively high. 2. Deliberate simplification for the benefit of the reader; seldom, if ever, necessary.

Engineer or scientist *To:* Upper echelon administrative official	information he needs to keep up to date, to manage and plan programs, to use in speeches and papers, or other. 2. To show that completed projects were justified; to justify new projects or continuation of old ones.	cial with technical training and/or special previous interest in the particular subject matter. 2. Very meager for officials without one of the above. 3. Wide variation between the extremes of Nos. 1 and 2.	that of the assumed extent of background knowledge—from relatively high to greatly simplified. 1. Relatively high. 2. No need to avoid technical terminology or to strive for simplification.
From: Engineer or scientist *To:* Fellow engineer or scientist	1. Usually simply to inform. 2. Sometimes also to obtain critical opinions from peers.	1. Extensive for readers working in same or closely related field. 2. Less extensive for a general reader group; good general technical knowledge can be assumed.	

13

obviously vary widely for different situations. If you are reporting experimental work you did yourself, most of the data will come from your own notebooks and other records. If your report involves comparing your new findings with previous work by yourself or others, you will need the papers and reports that give these results. Assembling information for whatever background discussion you consider appropriate may take you to many sources, including earlier reports, preprints, published papers, books, encyclopedias, and correspondence. You will find it helpful later if, during this roundup of raw material, you begin to think about how and where tables, graphs, drawings, and photographs can increase the effectiveness of your story. (See also Chapter 4.)

Ideally, at the end of this step, you have assembled, ready for use, every bit of new data you wish to present and every item of background information you believe you should give the reader. Just as with the amplifier builder, you probably will find as you write that you have forgotten a few essential "ingredients," and have assembled some others you don't need. Most of what you require will be on hand, however, and the closer you are to the ideal situation at this point, the more efficient your writing will be and the quicker you will finish your report.

PREPARING THE WORKING OUTLINE

Most of us have read (and probably written) papers and reports about which one or more of the following statements could be made:

1. Some of the information presented in the middle or toward the end of the document would have been more useful to the reader if it had appeared near the beginning.

2. Omission of significant pertinent material resulted in annoying gaps in the argument.

3. Some information was included which appeared to have no bearing whatever on the subject.

4. A few points were "beaten to death" by tiresome repetition in different parts of the paper.

The somewhat inelegant adjective "sloppy" probably best describes this kind of writing and the thinking behind it. Such a report is difficult to follow, commands neither the respect nor the interest of the reader, and cannot possibly perform its mission effectively. A carefully prepared outline, used as a writing guide, can go far toward saving you from committing this particular writing sin.

Some fundamental principles and practices of good outlining are discussed below. Keep in mind that at this point we are concerned *only* with an outline for you to follow while you are writing your report. Its relationship to the document's Table of Contents is covered in Chapter 6.

Selection of Heads and Subheads

Development of a good working outline begins with establishment of the main divisions of your report, and of the major heads to be used to identify them. If you follow the basic overall pattern mentioned in

Chapter 1, some of these heads will already be set up—Introduction and Conclusions, for example. In this case, you will be concerned primarily with how to break down the composite of several sections that were referred to in Chapter 1 as the Body of the Report. If, on the other hand, you decide upon a totally different basic approach, you will need to establish a scheme of major subdivisions for the report as a whole.

At first thought, several ways of organizing your material may seem equally logical. Consider, for example, a report on the study of reactions of the general public to advertising. Basic patterns for such a document could include:

1. Major divisions of people grouped into Women, Men, Girls, and Boys, with subdivisions under each for the various advertising media (e.g., radio, TV, newspapers, magazines, skywriting, billboards, bus cards, and bumper stickers) *OR*

2. Major divisions of people defined on any one of various other bases (e.g., age, profession, nationality, geographic location), with subdivisions under each for every advertising medium *OR*

3. Major divisions of advertising media (as suggested in No. 1), with subdivisions under each for different groups of people

Although all three patterns might be equally logical, one pattern would almost certainly be found preferable to the others for conveying the particular message and emphasis the author has in mind.

You may have some general ideas about main heads for your report even before you begin the pre-writing routine described in this chapter. By the time you have analyzed the problem and have assembled your data, you undoubtedly will have fairly clearly in mind at least one basic pattern. Write down this array of main heads, along with any alternative schemes that occur to you. Study them all, trying out different arrangements of subheads under each. Then select the one that you believe will best accommodate the information you wish to present and will most effectively convey the emphasis you wish to bring out.

In expanding your working outline under the pattern of main heads you have selected, carry the outlining to as many subordinate levels as possible. The greater the detail of your working outline, providing it is logical of course, the more valuable it will be as a writing guide. Subheads and sub-subheads that prove unnecessary as the actual writing progresses can easily be lopped off; adding new ones in the mid-writing stage is difficult and is likely to interfere with the logic of the overall development.

Equality and Subordination in Heads

Inherent in the very meaning of outlining is the concept of breaking a body of information into divisions and subdivisions among which exist certain recognized relationships of equality and subordination. Some basic principles underlying these relationships are given below.

BASIC PRINCIPLES OF REPORT OUTLINING

1. The main heads, taken as a whole, must be able to accommodate the total information to be presented—that is, every item of data must have a logical major-division home.

(For example, in the first alternative structure for the hypothetical people-and-advertising report mentioned above [page 16], the three main heads, Women, Men, and Girls, would be insufficient since there would be no place for recording the reactions of boys.)

2. The main heads must be consistent with each other and must not overlap.

(In the people-and-advertising report, the main heads, Women, Lawyers, Brunettes, and Homeowners, obviously would be unsatisfactory and would violate both of these precepts, as well as the first principle.)

3. No head must appear either equal to one to which it is logically subordinate, or subordinate to one to which it is logically equal.

(Men and Reaction to TV Commercials could not both be main heads in our illustrative report; depending upon the basic structural pattern of the outline, either might be subordinate to the other.)

4. There must not be a single main head.

(Such a head should be either the title of the report—or a part of it—or must have fellow main heads if all of the information is to be accommodated.)

5. Seldom should there be a single subhead.

(The reasoning here is the same as that given in No. 4. However, if maintaining parallelism [see below] is not important, an exception to this rule may be justified.)

Parallelism

Parallelism in an outline means repeating the same series of subheads under two or more successive heads at the next higher level. Parallel and non-parallel portions of the people-and-advertising outline might, for example, be as follows:

PARALLEL	NON-PARALLEL
I. Reaction to newspaper advertising	I. Reaction to newspaper advertising
A. By women	A. By women
B. By men	B. By men
C. By girls	C. By girls
D. By boys	D. By boys
II. Reaction to TV commercials	II. Reaction to TV commercials
A. By women	A. By blondes
B. By men	B. By brunettes
C. By girls	C. By red-heads
D. By boys	D. By baldheads

Parallelism that is consistent with the overall scheme of presentation offers obvious advantages in the development of a logical, easily read, effective report. Not all reports, however, lend themselves naturally to the use of parallelism in the organization of the information to be presented; in such cases you should not try to force the material into this kind of pattern.

Labelling the Heads

Systems for designating heads fall into two principal categories, with minor variations within each. As illustrated below, one of these categories alternates numerals (Roman and Arabic) with letters; the other follows a decimal pattern.

NUMBER & LETTER	DECIMAL
I	1
A	1.1
1	1.1.1
a	1.1.1.1
(1)	etc.
(a)	

Neither system is inherently better or worse than the other. In any given case the choice can simply be one of personal preference. Many people find the number-and-letter pattern a little easier to follow; it becomes a bit unwieldy, however, beyond the six levels of heading shown in the illustration. Then one gets into such labels as [(1)] and ([(1)]). Although the decimal system does not suffer as much from this shortcoming, it presents problems if more than nine items must be accommodated within any one subsection; the name decimal hardly fits, for example, a designation like 3.11.15. Actually, technical reports seldom require more than six levels of head or more than nine items under any one head. Whatever system you use, you must, of course, apply it consistently throughout any given report outline.

In summary, a detailed working outline is to your report roughly what a well-planned itinerary is to a successful automobile trip. In the latter case, you list in sequence all of the places you plan to visit. In a report outline, you list the points you wish to discuss in the order you have decided will tell your story most effectively. The traveller knows, of course, that no matter how carefully he plans his route, detours and unexpected points of interest will cause him to vary it a little as he goes along. Similarly, essential items of information that slipped your mind, and unforeseen weak points in your preliminary outline, will occur to you as you write and minor modifications will be needed. In neither the travel nor the technical writing case, however, does the fact that changes "en route" may be found desirable detract from the fundamental importance and value of preparing the itinerary or outline before you start.

SITTING AND THINKING

Although our discussion of this aspect of the pre-writing report preparation routine follows the others, in actual practice it is an exercise that should be carried out before and after each of the steps discussed above. There is no substitute for spending a certain amount of time quietly mulling over your report plans and letting a variety of ideas churn around in your mind. This kind of mental wrestling with the problems of analyzing, assembling, and organizing your report material frequently produces an almost magical effect. Often, ideas that may have seemed terribly confused and obscure suddenly sort themselves out and cause you to say to yourself, "Of course - how stupid of me; why didn't I think of that before?" But almost never does one think of these obvious solutions "before," and there probably is no short cut to the breaking of this particular kind of dawn.

CHAPTER **3**

Writing the Introduction

With this chapter we begin detailed consideration of the major sections of a representative technical report. The order in which we shall discuss these sections is one that you might logically follow in your writing:

1. Introduction
2. Body of the Report
3. Conclusions and Recommendations
4. Appendixes, Bibliography, and Table of Contents
5. Abstract
6. Title and Title Page

You will note that this sequence differs considerably from that shown in Chapter 1, where these same sections are listed in the order in which they might appear in the completed report. A point that is emphasized there is equally valid here. This preparation sequence, like the arrangement presented in Chapter 1, enjoys no divine stamp of approval as the only correct order. Although it is logical and can be followed effectively in most cases, you should not hesitate to vary it in situations where you are convinced a different sequence would be preferable.

The fundamental nature of the Introduction in a technical report has been well summarized by J. Raleigh Nelson[1] as follows:

The Introduction may be thought of as a kind of preliminary conference in which the writer and the prospective reader "go into a huddle" and agree in advance on the exact limits of their subject, the terms in which it will be discussed, the angle from which to attack it, and the plan of treatment that will be most convenient for them both.

To be consistent with this basic picture, we shall discuss the job of writing the Introduction, and then of the other sections, in terms of principal functions, format, and other considerations.

PRINCIPAL FUNCTIONS

Achievement of the objectives implied by the above quotation requires, as a minimum, introducing the reader to (a) the subject matter of your report, (b) your reasons for writing it, and (c) the plan you have followed in presenting its contents.

[1]Nelson, J. Raleigh, *Writing the Technical Report*, McGraw-Hill, New York, 1952, p. 36.

Subject and Scope

Top priority among these functions unquestionably goes to letting the reader know exactly what your report is all about. He should encounter this information very early in the Introduction—ordinarily within the first two or three sentences. Since this objective is also that of the ideal title (see Chapter 8), this portion of the Introduction can be thought of as an amplification of the title's highly condensed message. Often it is desirable to repeat the title in the Introduction's statement of the subject of the report. Along with informing the reader of what the report concerns, you should let him know the general limits of the coverage—that is, what to expect in terms of where the "story" begins and ends. We refer here to matters such as ranges of parameters dealt with, time period covered, whether the investigation being reported was experimental or theoretical, and the like. State such information as quantitatively as possible. For example, do not write "high temperatures" or "recent work" or "northern latitudes" if you can state precise temperature ranges, dates, and regions.

Purpose

Underlying every technical report is the purpose the author had in writing it—the element that makes it different from every other such document on the same subject. Note that here we mean the purpose of the report, not that of the experiment, theory, or device being reported.

Think of your report as a document with a definite

job to perform. It is intended to do something for someone, and that someone's needs are the determining considerations for your report. For example, the purpose of a report on blood transfusion probably would differ considerably depending upon whether it was being written for potential donors, nurses assigned to blood centers, or manufacturers of medical equipment.

In summary, if the reader is to learn from your report what he is supposed to learn, and what you want him to learn, he should be informed before he starts the report proper about what you propose to do for him, what emphasis you expect to maintain, and what point of view you are presenting. One author has aptly described the purpose of the technical report as the writer's answer to the hypothetical reader inquiry, "Why is this report important enough for you and me to spend time on it?".[2] Make certain that the Introduction in your report answers this question for your readers.

Plan

The Introduction should acquaint your reader with the general plan of presentation followed in the report. Sometimes this responsibility can be adequately discharged by a single sentence, such as, "In this report on bourbon-powered rockets, the data and recommendations are discussed under the following chapter headings: - - -." Seldom will more than two or three sentences be necessary or desirable. The objec-

---

[2] *The D(ratted) P(rogress) Report*, Savannah River Laboratory, E. I. du Pont de Nemours, Augusta, Florida, 1954, p. 7.

tive is simply to orient the reader of your report to its structure before he embarks upon the main body of the document—in short, to provide him with information about your report analogous to that which a tourist gets from his tour map, or a concert-goer from the program of the evening's performance.

The three functions discussed above should be performed by the Introduction, or introductory paragraphs, of every technical report. Other roles that the Introduction *may* also play, depending upon the particular reporting situation, are mentioned later in this chapter.

FORMAT

For purposes of discussion, we assumed above that your report would have a specific section or chapter labelled Introduction. Actually, the optimum format for the introductory material varies widely, depending upon the nature, complexity, and length of the document. Among the possibilities, beginning with the simplest, are:

1. One or two introductory paragraphs without any specific heading of Introduction—used appropriately when the report is short

2. A section headed Introduction, but with no subheads—suitable for a report somewhat longer than that referred to in No. 1, but still neither very long nor complex

3. A section or chapter headed Introduction, with two or more subsections carrying subheads—typically employed in fairly lengthy, elaborate reports

4. No actual heading of Introduction but several major heads of equal stature under which are presented the several kinds of information described in this chapter as introductory in nature.

Regardless of which format you choose, make certain your introductory paragraphs fulfill the three principal functions relating to subject and scope, purpose, and plan, as well as any other item mentioned below which you deem appropriate in your report.

OTHER CONSIDERATIONS

An imaginative, well-constructed first sentence (and paragraph) can do much to orient the reader properly and put him in a comfortable, receptive mood for the main body of your report. Conversely, a drab, unimaginative, poorly phrased initial sentence may cause him to read no further, or at least to do so with reluctance and prejudice. Two ways *not* to start the Introduction are:

1. With chronology or history—although you may wish to include some information of this kind elsewhere in the introductory paragraphs (see below) *OR*
2. With a rambling general discussion which approaches the topic through a series of approximations and keeps the reader guessing for several paragraphs as to just what the report is all about

Initial sentences and paragraphs of these kinds usually result from a combination of simple laziness, inadequate pre-writing study and meditation, and an excess of impatience to start putting words on paper. You should let the reader know the subject of your report immediately, and try to make the very first sentence one that will encourage (indeed, entice) him to read on. Before you start your Introduction, study the first paragraphs of a number of reports, noting particularly their "kick-off" sentences. From these you undoubtedly will get ideas both on how to start and how not to start your report.

The possibility of incorporating historical material in the Introduction, or introductory paragraphs, was mentioned above. First, you must decide whether such information is to be included at all. If you believe some historical background will help the reader get your message, and thereby contribute to the document's effectiveness, by all means include it; if not, omit history. In most cases, you will probably think that at least a little historical information is desirable. If your objective in this regard can be taken care of with a few sentences, the Introduction is usually the most appropriate place for them. If, however, you wish to present a fairly extensive historical discussion, a separate section generally is preferable. This alternative is discussed in Chapter 4.

In technical reports on experimental investigations, the situation with regard to theoretical background or analysis is much like that just described for historical background. If a few sentences will suffice, incorporate them in the Introduction; if effective presentation of the message requires something more extensive, set up a separate section or sections, as outlined in Chapter 4.

A few suggestions have already been made regarding the optimum sequence for the several elements of the Introduction. If your Introduction (or introductory paragraphs) is to perform only the three principal functions, the sequence should be that in which they were discussed earlier in this chapter—subject and scope, purpose of the report, and plan of presentation. If additional matters are to be covered in the Introduction, they should usually be treated after the statement on purpose. An exception may be desirable, however, in the case of historical background that is to be included in the Introduction. Sandwiching the two or three sentences on history between an initial sentence or two on the subject of the report and some expansion of subject and scope can often be very effective. In any case, the statement on what the report is about should come at the very beginning of the Introduction, and the statement on the plan of presentation at the end.

CHAPTER 4

Writing the Body of the Report

The term body of the report, as used here, includes the several separately named sections that constitute what may be described as the "meat" of the document; it never appears as an actual section heading. The basic function of these sections is to answer three questions for your readers:

1. What did you do?
2. How did you do it?
3. What did you find out?

Depending upon the particular reporting situation, adequate answers to these queries may require including certain historical, theoretical, or other background information as well. These body-of-the-report sections typically will appear in the finished document between the Introduction and the Conclusions. Note that they do not *include* your conclusions and recommendations. What you conclude from your results, and what you may wish to recommend, both depend upon, and stem from, your answer to question No. 3, but neither is strictly a part of that answer. (Writing the conclusions and recommendations is discussed in Chapter 5.)

The body of the report ordinarily constitutes the bulk of the document. Its sections require less discussion than some of the others, however, because the nature of each is reasonably clear from its name. The optimum assortment of such sections varies so greatly from report to report that no single grouping or sequence can be identified either as representative or as warranting recommendation for use in the majority of cases. As you plan this portion of your report, however, there are certain basic considerations you should have in mind. Passing reference was made to some of these in Chapters 1 and 2. They are considered further in the rest of this chapter as they relate, first, to the kinds of material appropriate for the main-body sections, and, second, to the most effective pattern of presentation. In preparing the detailed working outline described in Chapter 2, you will already have developed a first approximation to such a pattern; you should now refine it in the light of the suggestions given below.

KINDS OF MATERIAL

The following listing of several groups of actual section heads is taken from the main-body portions of a number of randomly selected technical reports:

- Historical Summary; Historical Background; History

- Description of Equipment; Description of Apparatus

- Design of Equipment; Design of Apparatus

- Theoretical Considerations; Theoretical Background; Theory

- Predicted Performance; Observed Performance

- Data; Working Data; Computations; Computations and Data

- Results; Computations and Results; Summary of Results

Roughly similar heads are grouped together in this listing. Although no one technical report is likely to require headings representative of all of these categories, and many other combinations and choices of words are possible, the actual section titles in this tabulation are reasonably typical, and do suggest most of the *kinds* of information suitable for inclusion in the body of the report.

How should you go about selecting the particular mix of main-body sections that will be optimum for your report? First, try to put yourself in the position of a prospective reader. Then, review carefully, and modify where necessary, your earlier answers to the four questions posed in Chapter 2 under "Analyzing the Overall Problem." Make certain that "you, the reader" are fully in agreement with the decisions of "you, the writer" regarding who your readers will be, what you want to tell them, and what you can assume they already know. Now you are ready to resume the role of author and to decide what sections should comprise the body of your report.

Data and results (if the work being reported upon was experimental) or their analogues (for a totally theoretical treatment) obviously must be presented since, regardless of the readership, these are what the report is all about. Whether you should also include material on history and on theoretical or experimental background—and, if so, how much—depends largely upon the extent of background knowledge you can assume your readers will possess. If all, or almost all, potential readers can be presumed to be fairly expert in the subject area involved, a minimum of such material is necessary. But if the report's probable readership includes a number of individuals not particularly knowledgeable in the specific subject field of the report, even though they may be broadly competent in science or engineering, more extensive supplementary information is called for.

With regard to background information in general, erring a bit on the side of telling the reader more than he might absolutely have to know is preferable to taking any chance on not telling him enough. After all, potential reader groups seldom can be defined with great precision; individuals within such groups vary considerably in their background knowledge;

and most worthwhile technical reports prove to be of some interest to at least a few people who do not belong to what the author may think of as his report's prime public. The specific nature and objectives of the report must also be considered when you are evaluating the need and desirability of providing historical and theoretical background information.

Among the other kinds of supplementary information suggested by the earlier listing of actual main-body sections are design and description of equipment or apparatus, and data on predicted versus observed performance. As already noted, that list is illustrative, not comprehensive, and you may think of other items that should be considered for inclusion in this portion of your report. In judging each of these items, you should decide whether it is (a) essential to the document's mission, (b) not essential but perhaps useful, (c) harmless but of no real value, or (d) completely out of place. Then act accordingly. As with the other items, the nature of the anticipated audience will have an important bearing on your decision. Of greater significance in this case, however, are the nature of your report and your objectives in writing it.

Other important main-body decisions you must make relate to the use of tables, drawings, graphs, and photographs. Our discussion here of these elements is limited to brief consideration of one or two basic principles.[1] The visual aids that accompany technical

[1] A number of the publications cited in Appendix B discuss the actual preparation of tabular and illustrative material.

report exposition are ordinarily intended to accomplish one or both of two objectives: to make the technical message more effective, and to add to the impressiveness of the report's appearance (the keeping-up-with-the-Joneses syndrome). It is when concern with the latter objective has been excessive that one finds fancy drawings in more colors than are technically warranted, excessive use of overlays, fetching photographs of glamorous typists apparently operating complex scientific equipment, and the like—in short, illustrations with much closer ties to the public relations office than to the laboratory. For guidelines on the selection and presentation of illustrative material intended to serve this second function, you are on your own as far as this book is concerned. While this objective is not totally unworthy, methods for achieving it are mostly based on other than scientific considerations and, therefore, are not appropriate for discussion here.

Apropos of the first objective—improving the technical presentation—the cliché that a picture is worth X thousand words is familiar to everyone. Unquestionably, words plus illustrations frequently can put ideas across more quickly, accurately, and efficiently than can words alone, and tabular presentations of data often are superior to straight narration. It is also true, however, that in statements like these the significance of such qualifying words as "frequently" and "often" is not always sufficiently recognized. A picture is not worth X thousand words if X+1 thousand words are required to explain it; data do not always lend themselves to tabular or graphical presentation; poorly designed and badly reproduced graphs and drawings may be more confusing than enlightening; color in illustrations is not necessarily better than black and white, nor are seven colors always preferable to two or three; a graph may or may not be more effective than a table; and so forth.

So, the problem that you as a report author face with regard to tables and figures is exactly the same as the one that confronted you in the selection of the kinds of information to include in your report. Again, you must apply the yardstick of maximum effectiveness from the standpoint of your report's readers. Use tables, graphs, drawings, and photographs when (and only when) you believe they will benefit the reader.

SEQUENCE OF SECTIONS

Since the optimum assortment of sections for the body of a technical report varies greatly with the reporting situation, no tidy, always-applicable recipe can be stated for arranging them. Once more, the fundamental consideration must be maximum service to the reader. Select a pattern that will lead your reader logically from the first section into the second, from the second logically into the third, and so forth. Do not inflict upon him the report-reading equivalent of driving from New York to Pittsburgh by way of Kansas City. Present the various sections in the order you believe will make the best sense to him. One or two general ground rules are fairly obvious. If you are including material on history, the best place for it is usually at, or very near, the beginning of the body of the report. Theoretical background or information on design of apparatus ordinarily should precede presentation of data and results that stemmed from that theory, or were obtained with that apparatus; on occasion, it might be better to place such material in an appendix. (See Chapter 6.) Establishing a logical sequence for the main-body sections is seldom difficult; the principal point we wish to emphasize is simply that you should give careful thought to doing it.

The report that has to be directed both to technical experts (for example, the director of research or a fellow scientist or engineer in the same field) and to non-experts (for example, the president of the company or the commanding officer) presents a special dilemma. Extensive background will bore the expert before he gets to the data and results; without substantial supplementary material, the non-expert will probably miss the message. A compromise approach that frequently works very well in such cases is to insert after the Introduction a kind of once-over-lightly account that presents in non-technical or semi-technical language both the necessary background and a summary of data and results adequate for the non-expert. This account can then be followed with the full detailed treatment for the expert. The reader who is scientifically competent and well informed in the field can omit the semi-technical material and concentrate on the detailed sections; the non-expert can do the reverse.

Finally, in completing the body of the report, check back over both your selection of sections and the order in which you have arranged them, and ask yourself whether they do indeed answer the three questions of what you did, how you did it, and what you found out. If they do, and if your presentation is effective, the reader will be properly informed and oriented to proceed intelligently to perusal of your conclusions and, if any, recommendations.

CHAPTER 5

Writing the Conclusions and Recommendations

Every good technical report has an introduction, regardless of whether the introductory sentences or paragraphs are so captioned. Similarly, every technical report has a body of the report although, as discussed in Chapter 4, the group of sections encompassed by this label varies widely for different reporting situations. But not every technical report contains conclusions and recommendations; also, some reports may include the former, but not the latter. A progress report, for example, may consist simply of straightforward exposition of the what-we-did-last-month kind, and may neither conclude nor recommend anything. (Remember that in the wonderful world of technical-report jargon, progress includes nonprogress.) A report on an experimental task or project usually (but not always) contains conclusions; it may or may not also present the author's recommendations for further action of one kind or another.

This chapter is concerned with the kind of technical report that includes *both* conclusions and recommendations. These may appear as separately labelled sections, or they may be combined in a single section. When conclusions and recommendations are present, they usually constitute the climax of the report—the principal reason it was written. They are discussed below with respect to principles of preparation, acceptable format, and relation to the general report structure.

PRINCIPLES OF PREPARATION

Two basic defining distinctions must be made clear at the beginning of this discussion. One distinction is between conclusions and conclusion; the other, between conclusions and recommendations. The former difference is expressed concisely by Sherman in his statement, "Conclusions are 'convictions arrived at on the basis of evidence' rather than just 'the section that comes at the end.'"[1] In the case of a technical report, the "evidence" is contained, of course, in what we have called the body of the report.

Although one might think the difference between conclusions and recommendations would be clear from their names, in practice the two are frequently confused. As noted above, conclusions are derived from known data and concern the present. Recommendations, on the other hand, propose that some-

---

[1] Sherman, Theodore A., *Modern Technical Writing*, Prentice-Hall, Englewood Cliffs, N.J., 1966, p. 197.

thing be done, or not be done, in the future. To illustrate, Joe's behavior at the office Christmas party provides data from which his boss draws certain *conclusions*. The several *recommendations* his boss then makes regarding Joe's future conduct stem from, but are not identical with, these conclusions.

Two fundamental principles must be observed if the conclusions you present in your report are to be valid and appropriate. One is implicit in Sherman's statement quoted above, namely, that every conclusion you state in this section of your report should be based on information that appears in the body of the report. Further, the line of reasoning that took you from one to the other should be made very clear to the reader. "Floating" conclusions for which no background evidence appears in the report are completely out of place, regardless of how true you may believe them to be, or how emotionally committed you may feel to the sentiments they express. Either document them, or omit them.

The other fundamental principle relating to conclusions concerns how your statement of the conclusions compares with what you wrote in the Introduction. If you follow the precepts outlined in Chapter 3, your draft introduction will be designed, in part, to orient your readers to a certain general pattern of development in the rest of the report. What you now conclude in this section should be compatible with the kind of result the Introduction promised—that is, within the "ball park" in which you have indicated the emphasis of the report will fall. Suppose, for example, that your report concerns a series of rocket tests and, in the Introduction, you state that the objective is to compare distances travelled for various sizes of fuel charges. Now, if your conclusions actually relate to relative costs of flights of a given length for different kinds of fuel, you will not have played fair with your readers and you will leave them in a state of confusion. So, check your tentative conclusions against the draft of your introduction, and if any of them appear out of line with what you led the reader to expect, do the necessary rewriting of the Introduction, of the Conclusions, or of both.

The inherent distinction between conclusions and recommendations has already been pointed out and illustrated. With minor word changes, substantially everything noted above about the conclusions can also be said of the recommendations. Just as every stated conclusion should have its roots in specific material in the body of the report, so every recommendation should stem directly and clearly from one or more of the conclusions. A recommendation with no visible means of support among the conclusions is as much out of place as a conclusion which cannot be readily traced to the data and results. Similarly, what you recommend should be harmonious with the general scheme to which you oriented the reader in the Introduction. Make no recommendation that is merely a helpful suggestion that happens to occur to you, but is not directly related to the conclusions of your report, or that is an unsupportable plea for some kind of incidental course of action you favor. As with the conclusions, either document your recommendations, or omit them.

FORMAT

If your report is short, and your conclusions few and relatively obvious, you may wish to present them in a brief expository paragraph. If you plan to make no more than one or two recommendations, the same can be said of them. Whether you should have separate specific headings of Conclusions and Recommendations, a composite section titled Conclusions and Recommendations, or merely an unlabelled paragraph or two stating what you conclude and recommend, depends upon how you have handled the head-and-subhead problem in the rest of the report.

On the other hand, if your report is somewhat long and fairly complex, and if you wish to present more than two or three conclusions, or conclusions and recommendations, list and number them consecutively. Again, depending upon their number and complexity, you may wish to have separate sections headed Conclusions and Recommendations, or a single section that combines both. Each conclusion and recommendation so listed should be accompanied by the argument which links it to information in the body of the report (if it is a conclusion) or to one or more of the conclusions (if it is a recommendation).

LOCATION IN THE REPORT

In the strictly conventional sequence of sections in a technical report, the section Conclusions immediately follows the body of the document, and Recommendations follows Conclusions. (If the two are combined in a single section, it of course follows the body of the report.) For a report written according to this pattern, the reader is first oriented to what the report is about and to why it was written (Introduction); second, he is informed about procedures and findings (Body of the Report); third, he learns what the author concludes from these results (Conclusions); and, finally, he is acquainted with what the author thinks ought to be done about it (Recommendations). This pattern has been called the inductive plan, and certainly it takes the reader from point of departure to destination along a tidy, logical, straightforward route.

There is much to be said, however, for adding one small detour to this straight-and-narrow reporting path—a digression called Summary of Conclusions and Recommendations in the representative list of parts of a technical report given in Chapter 1. (In a report that makes no recommendations, this section would simply be headed Summary of Conclusions.) In that listing, the summary was placed immediately after the title page. To state conclusions and recommendations at this point admittedly is a little like revealing the name of the murderer in the first page of a detective story. In a mystery tale this would be very bad form. Maintaining suspense, however, is not a proper objective in technical report writing. The fundamental goal is effective communication of technical information, and an early summary of this kind can often contribute a great deal to its accomplishment. Inclusion of such a summary is advisable in a report that must serve both management and the author's fellow engineers or scientists. The administrative official who wants to know the "answers" first can find them readily. Then, at his leisure, he can read the complete account, and study the procedures, data, results, and the reasoning that led to the conclusions and recommendations. Frequently, the practicing scientist or engineer also finds it helpful to have the conclusions and recommendations in mind as he reads the body of the report.

Do not, however, think of this early summary as in any sense replacing the complete section, or sections, on these topics. The summary should list only the specific conclusions, or conclusions and recommendations, expressed as concisely as possible and uncluttered by any background material whatsoever. To introduce supporting evidence at this point would remove much of the summary's ready-reference value and defeat the real purpose of including it.

CHAPTER 6

Preparing the Appendixes, References, and Contents

Grouping appendixes, references, and contents for discussion here is not meant to imply an organic relationship or interdependence among them of the kind discussed in the preceding chapter for conclusions and recommendations. These sections really have in common only that they are all supplementary to the main business of the report, and can best be written, at least in final form, after the principal message-carrying parts of the document have been completed.

APPENDIXES

Like some other sections of a technical report, the element of appendixes does not lend itself to simple, precise definition. Probably the most satisfactory approach to consideration of this section is via the general-definition and representative-example route.

General Definitions and Examples

The following quotations relate to the general nature and appropriate functions of technical report appendixes, and suggest the basic frame of reference within which you should draft this section of your report—if you decide appendixes are called for:

1. Appendixes present "Explanatory material which cannot be given in the text of the report without interfering with the logical and orderly progress of the paper." [1]

2. "An appendix relieves the body of the report from congestion. It presents pertinent data that are too detailed to be given in the text." [2]

3. Appendixes "represent the section into which all material not directly contributory to the reader's understanding of the subject is placed." [3]

4. "The appendix contains supporting material—material not critical to the understanding of the text but which is of interest (and help) to some readers." [4]

Some of the specific kinds of information found in a sizable random sample of actual technical reports are listed below. Keep in mind that these examples

---

[1] Crouch, W. G. and R. L. Zetler, *Guide to Technical Writing*, Ronald Press, New York, 3rd ed., 1964, p. 126.

[2] Nelson, J. Raleigh, *Writing the Technical Report*, McGraw-Hill, New York, 1947, p. 284.

[3] Graves, H. F. and L.S.S. Hoffman, *Report Writing*, Prentice-Hall, Englewood Cliffs, N.J., 4th ed., 1965, p. 129.

[4] Rathbone, Robert R., *Communicating Technical Information*, Addison-Wesley, Reading, Mass., 1966, p. 74.

are meant to be representative, not comprehensive, and that the qualifying condition *not essential to the report's main argument* is to apply in all cases.

1. Supplementary calculations that might be of interest to some readers.

2. Derivations of formulas used in the report, when these are sufficiently specialized or unusual to be of interest to readers.

3. Auxiliary charts, maps, graphs, and the like that may help clarify the principal data presented in the main body of the report.

4. Complete tables if these have only been summarized in the report proper, and the author believes the reader should have the details available.

5. Pertinent data too detailed for inclusion in the main body of the report.

6. Supporting data and computations, or other material, useful to the reader who wishes to confirm the report's findings.

7. Descriptions of experimental or theoretical approaches that failed before the one described in the report proper was tried.

8. Bibliographies or lists of references. (See the next major section of this chapter.)

These examples and the broad descriptive quotations that preceded them should enable you to decide whether your report needs an appendix and, if so, what material presented in this manner will add to the overall usefulness and effectiveness of the document.

Format

The optimum pattern of presentation for any given appendix varies with the nature of the material being presented. The best plan may be simple exposition, tables or graphs, an outline, one or more lists, or some other format. In the report as a whole, the appendixes usually appear as the last, or next-to-last, section. The latter arrangement is used in the representative scheme shown in Chapter 1.

REFERENCES AND FOOTNOTES

These report elements are considered together here because both frequently involve the citation of other publications, and, when they do, they should employ the same form of entry.

References

Other publications are usually cited by name in a technical report for one or both of two reasons: either the documents have been specifically referred to in the text, or the author believes they provide worthwhile supplementary information on the subject of the report. Publications of the former type may be grouped under a heading of References or List of References. Lists of items of the supplementary kind are more likely to be labelled Bibliography or Supplementary Reading. In technical report literature as a whole, however, the distinction between "specifically referred to" and "supplementary only" is not followed very strictly, and the two kinds of reference are often intermingled, with the combination labelled either References or Bibliography. In any of these lists, the entries may be simple citations or they may include brief descriptive annotations.

If you examine the lists of references in a number of publications chosen at random, you will find variations in the forms of entry. The basic paraphernalia of capital letters, small letters, italics, boldface, abbreviations, underlining, commas, colons, semicolons, and periods appear in a variety of combinations. Although individual writers sometimes defend one or another of these schemes with deep emotional fervor, the fact is that no one pattern is any more right or wrong than another, as long as all of the essential data are included. Another important requirement is to be consistent, at least within any one report, and preferably throughout all of the reports issued by a given laboratory. Your organization may require a particular style of citation. If so, this problem is solved for you. If not, select one from the many style manuals that are available. (See Appendix A.) One acceptable pattern is as follows:

1. Book or monograph with single personal author.

   a. Sequence of elements: Author (last name first). Title (underlined). Publisher, location of publisher (city only for major cities), Edition (if other than the 1st), Year of publication. Number of pages, or specific page reference, whichever is most appropriate.

167901

b. Example: Nelson, J. Raleigh. Writing the Technical Report. McGraw-Hill, New York, 2nd ed., 1952. 355 p.

2. Book or monograph with more than one personal author.

a. Sequence of elements: Authors (cited alphabetically, with last name first for initial author only). Same as 1-a for elements after title.

b. Example: Graves, H. F. and L.S.S. Hoffman. Report Writing. Prentice-Hall, Englewood Cliffs, N.J., 4th ed., 1965. 286 p.

3. Book or monograph with no personal author.

a. Sequence of elements: Title first. Same as 1-a for elements after title.

b. Example: American Heritage Dictionary. American Heritage Publishing Co. and Houghton Mifflin, New York, 1969. 1550 p.

4. Journal article with single personal author.

a. Sequence of elements: Author (last name first). Title of article (in quotes). Name of journal (underlined), volume (underlined), issue number, page numbers, date of journal issue.

b. Example: Gray, Dwight E. "Technical Reports I Have Read—and Probably Written." Physics Today, 13, 11, 24-32, November 1960.

5. Journal article with more than one personal author.

a. Sequence of elements: Authors (cited alphabetically, with last name first for initial author only). Same as 4-a for elements after title.

b. Example: Futrell, J. H. and T. O. Tiernan. "Ion-Molecule Reactions." Science, 162, 3852, 415-422, 25 October 1968.

6. Journal article with no personal author.

a. Sequence of elements: Title first. Same as 4-a for elements after title.

b. Example: "Wasted Words; Languages of Science." Scientific American, 194, 4, 71, April 1956.

7. Technical reports.

a. Sequence of elements: In many, perhaps most, cases of technical report citation, the corporate author (i.e., the issuing agency) is of greater interest and significance than the personal author. Depending upon the particular situation, therefore, the optimum arrangement of elements in the reference may be *EITHER*

(1) Corporate author, location of corporate author, title of report (underlined), personal author(s), date, number of pages, identifying numbers that appear on cover or title page (in parentheses) *OR*

(2) Personal author or authors (as described in 2-a and 5-a), title of report (underlined), corporate author, location of corporate author, date, number of pages, identifying numbers that appear on cover or title page (in parentheses)

b. Examples of each arrangement.

(1) Sandia Labs., Albuquerque, N.M., A Lead Zirconate Titanate Stress Transducer, R. E. Hutchison, July 1969, 27 p. (SC-DR-69-356)

(2) Likins, P. W., Dynamics and Control of Flexible Space Vehicles, Jet Propulsion Lab., Calif.

Inst. of Tech., Pasadena, 15 February 1969, 90 p. (NASA-CR-105592; JPL-TR-32-1329)

Footnotes

Technical reports do not normally require many footnotes and, for ease in reading, it is desirable to keep the number to a minimum. In a few situations, however, a footnote is clearly the best way to provide the reader with a bit of significant associated or supplementary information that does not fit in the text proper. Three functions that footnotes can serve effectively are outlined below, together with some comments on each.

1. To provide the exact source of, and give credit for, data or quotations taken from someone else's work or publication. Unacknowledged use of another person's writing is the kind of stealing we call plagiarism. Identifying such sources can add appreciably to the prestige of your report—particularly if they are widely recognized as authoritative.

2. To provide limited supplementary information on a particular point—information that might distract the reader if it is included in the text, but which you believe he should have. This same kind of function was mentioned earlier, you will recall, as an appropriate role for an appendix. The choice between footnote and appendix depends largely upon the amount of material you wish to present. If this amount is substantial (for example, a long derivation of an equation used in the report, or extensive supporting computations), an appendix is clearly indicated. If, how-

ever, the supplementary information can be covered
in a sentence or two, a footnote should be used.

3. To provide cross references to other parts of
your report. Footnotes of this kind may refer either
to preceding or following sections of the main text or
to the appendixes. They help assure that the reader
will view a given point in the proper context.

Ground rules for writing footnotes are no more
absolute and precise than are those for preparing
some of the other report sections we have discussed.
Again, optimum style and format vary with the reporting situation and, within broad limits, adherence
to a particular pattern is more important than the
scheme itself. The following general rules represent
one acceptable system.

1. Location of footnotes. From the reader's standpoint, the most convenient place for a footnote is the
one implied by its name, that is, at the foot of the
page (of the final report) on which the reference to it
occurs. This is the practice usually followed, except
when the reference is to a table or to a tabular entry;
in this case, the footnote usually appears at the bottom of the table. Sometimes, convenience of the
printer and publisher is given precedence over that of
the reader, and footnotes are grouped together on a
page at the end of the report; in this form they are
more properly called end-notes.

2. Identification of footnotes and reference
points. Superscript Arabic numbers or superscript asterisks are generally used at the points of reference to
indicate footnotes. The line spacing in most technical

reports permits such superscripts. But if they cannot be used, the footnote identification number can simply be inserted in parentheses at the appropriate place in the text. The index number or asterisk is placed at the point at which the reader is likely to wonder about the source, or the point at which you want him to have the footnote information—for example "... as the Hotchkiss hypothesis[1] suggests ..." or "... McGillicuddy's latest results* show that ..." For several footnotes on the same page, consecutive numbers or multiple asterisks are used; the asterisk scheme is not satisfactory, however, for a page with more than two footnotes. If the footnotes appear at the bottoms of pages throughout the report, they may be numbered either by pages or in a single consecutive series; the former pattern is probably more common. End-notes must be numbered in a single series.

3. Style of footnotes. For a footnote that simply identifies a published or unpublished source of material, use the citation style that you employ for other references to books, journal articles, and technical reports. (See preceding subsection on References.) Footnotes which are intra-report cross references may consist only of "See" or "See also" followed by page or section numbers, or they may also include short annotations. Footnotes that supply supplementary text information are phrased as the particular situation requires. Manuscript typing instructions for separating footnotes placed at the bottom of a page from the text material above are given in Appendix C on page 111.

TABLE OF CONTENTS

If your report exceeds two or three pages in length, and has more than a couple of sections with headings, it should have a Table of Contents, frequently called simply Contents. Such a table provides both a quick reference to the subjects dealt with in the report and a guide to the location of specific topics and subtopics.

Adapt your Table of Contents basically from the *final form* of your working outline—that is, from the original working outline (see Chapter 2), revised and modified to fit the completed text. This detailed scheme may extend to several decks or echelons. A Table of Contents for a technical report seldom needs more than two degrees of subhead, and often only one is necessary. Subordination is indicated by indentation just as in the working outline, but the numeral or letter designations of the subheads are frequently omitted. The number of the page on which each section or subsection starts should always be given.

Make sure that every item in the Table of Contents of your report appears as a head or subhead in the text itself. Every head or subhead in the text need not be included in the Table of Contents, however. Whether the Table of Contents should be given a page by itself, or be placed on the same page as the List of Illustrations or Figures, depends upon the lengths of both. If you put them on the same page, however, make certain both are complete on that page. Locate the Table of Contents immediately following the Title Page or the initial Summary of Conclusions and

Recommendations, if you have included such a summary. (See the representative sequence of report sections presented in Chapter 1.)

LIST OF TABLES AND LIST OF ILLUSTRATIONS

Little need be said about these lists beyond noting, first, that you should provide them if there are more than two or three labelled tables or figures and, second, that sometimes they are combined into a single list. The usual location for such lists is immediately following the Table of Contents. Be sure to give the number, exact title, and page location of each table or illustration.

CHAPTER 7

Writing the Abstract

We shall begin this chapter with two very general statements about the abstracts authors write for their technical reports. The first is a broad definition; the second is a dogmatic pronouncement.

1. An abstract of a technical report is a highly condensed, factual summary of certain information that appears in greater detail in the full document.
2. No self-respecting technical report more than a page or two in length should appear in public without an abstract.

Beyond these two generalizations lie a number of complexities that arise from the differing functions of, and criteria for preparing, abstracts of several different types. In this chapter, we shall define the two basic classes of abstract, relate the abstracts of the real world of technical reports to these "textbook" classes, mention briefly the place of critical reviews in the abstract picture, and, finally, consider principles and rules that pertain to the writing of effective technical report abstracts.

PRINCIPAL TYPES AND FUNCTIONS

By name, the two basic types are *descriptive* (also called *indicative*) abstracts, and *informative* abstracts. These labels themselves imply the primary functions of the two kinds of abstract with fair accuracy, but let us look at them in some detail.

Descriptive (or Indicative) Abstracts

A descriptive abstract describes, or indicates, the nature of the contents of the parent report, that is, it tells what the report is all about. If, for example, the work being reported was experimental, a descriptive abstract informs the reader briefly and concisely about such matters as what experiment was performed, how the task was approached, what kinds of data and results are presented in the report, and whether conclusions and recommendations are included. For a report on theoretical work, this kind of abstract indicates, as a minimum, the nature of the problem, how it was approached, and whether a solution is offered. In neither the experimental nor the theoretical case would a purely descriptive abstract present actual data, results, conclusions, or recommendations.

In short, in its pure form, a descriptive abstract is a kind of table of contents presented in narrative style. It can help the reader decide whether he wants to read the complete report. It cannot serve him as a substitute for the full account if he is interested in actual data and results.

Informative Abstracts

An informative abstract not only acquaints the reader with what the technical report is about, but it also summarizes the principal findings, conclusions, and recommendations that appear in the document. It is therefore more efficient than its descriptive cousin in helping the reader make up his mind about reading the entire report. And, since it summarizes data and results, an informative abstract can sometimes serve satisfactorily in lieu of the complete account, particularly if the abstract appears separately and the parent report is difficult or impossible to obtain.

In theatrical parlance, we may think of a descriptive abstract as analogous to the program, or playbill, which the usher hands you in the theater; it merely lists the cast, sets the scene, and tells you when and where the action takes place. An informative abstract, however, corresponds more to a synopsis which, in addition to presenting the descriptive information, summarizes the play's action, reveals the nasty fate that befalls the villain, and discloses who gets the girl at the end of the third act.

Real-Life Scientific Abstracts

The abstracts that actually appear in technical reports usually do not separate tidily into the two pure forms described above. And, indeed, even these theoretical classes are not mutually exclusive. As our definitions indicate, the basic characteristics of informative abstracts include those possessed by descriptive abstracts. Real-life abstracts scatter along a kind of continuous spectrum from "ultra-descriptive" to "infra-informative"—that is, from annotations a sentence or two in length that offer no substantive information to data-packed digests that border on being short papers. Any given report abstract may be predominantly descriptive or informative, but it is not likely to belong at either extreme of this spectrum.

The following pair of abstracts of the same report—one strongly descriptive, the other decidedly informative—is representative of abstracts found in technical reports, and illustrates the differences between the two kinds.

Title of Report: The Effect of Injection Pumps on Cold Starting of Diesel Engines

Descriptive Abstract: Results are presented of a series of cold-room tests on a diesel engine to determine the effect on starting time of (1) fuel quantity delivered at cranking speed, and (2) type of fuel injection pump used. The tests were made at a temperature of -10°F; engine and accessories were chilled to -10°F for at least 8 hours before the experiment began.

Informative Abstract: Cold weather tests were made to determine the effect on cold starting of the quantity of fuel injected at cranking speed, for two types of injection pump. The diesel engine of the energy cell-Lanova type that was used had 3.75-in. bore, 5-in. stroke, and 331-cu. in. displacement. The cold room was maintained at -10°F; engine, batteries, fuel and lubricating oils, and all equipment were chilled to -10°F for at least 8 hours before the engine was started. Best starting was obtained with 116 cu. mm., or 85% in excess of the fuel required for maximum power. Very poor starting was obtained with the lean setting of 34.7 cu. mm. Results with two different pumps showed that they must have good distribution, even at low cranking speeds. Unless each cylinder contributes its best power under these conditions, cranking speeds will not increase as rapidly as they should, and a longer starting time is required.

Among journals devoted entirely to abstracts, *Engineering Index*[1] is quite typical of those whose contents are basically in the descriptive abstract category. The abstracts in *Chemical Abstracts,*[2] on the other hand, fall well along toward the informative end of the abstract spectrum.

The Critical Abstract or Review

The critical abstract is written by someone other than the author of the document to which it refers. Therefore, you need not worry about this category as far as writing the abstract of your report is concerned. We mention it briefly here merely to complete our group portrait of the "abstract" family.

As its name implies, the critical abstract expresses a quality judgment. That is, in addition to information on the nature of the report or article and a short summary of the findings that the report presents, this kind of abstract includes the abstracter-reviewer's

[1]Published by Engineering Index, Incorporated.
[2]Published by the American Chemical Society.

opinion of the quality of the work that is reported, of the report itself, or of both. In terms of our theatrical analogy, critical abstracts correspond to dramatic reviews, and, in fact, often are called reviews. The extent to which such an abstract helps the reader decide how seriously he wishes to take the report depends, obviously, upon his own opinion of the knowledge and competence of the reviewer, or of the standing of the publication in which it appears. Typical of abstracting journals that carry critical abstracts is *Mathematical Reviews*.[3]

PREPARING YOUR ABSTRACT

The total process of preparing your abstract can be divided into two sequential stages: (1) deciding what type of abstract to write, and (2) writing it.

Choosing the Type of Abstract

Before you can start writing your abstract, you must decide exactly where you wish to peg it along the spectrum of informativeness described above. Do you believe the abstract of your report should be purely descriptive, heavily informative, or something in between? If your laboratory or agency has specific regulations regarding the nature and length of the abstracts that accompany the reports it issues, this decision has already been made for you. Or you may be required, or at least find it advisable, to follow the abstracting policy of some general abstracting and

[3] Published by the American Mathematical Society.

indexing service that covers your organization's reports. If, however, this choice is one you must make for yourself, your fundamental objective, as always, should be to make the abstract of maximum usefulness to those who will comprise your principal reader group.

As a semi-generalization, we can say that an appreciable degree of informativeness is desirable, especially in abstracts of reports on specific research tasks or projects. A study of physics abstracting[4] made some years ago disclosed that research physicists expressed an overwhelming preference for informative rather than descriptive abstracts. These same physicists also agreed almost unanimously that they would never use even an informative abstract as a substitute for the parent report if the complete document were available. Actually, the two views are not incompatible. These physicists simply believed very strongly that *only* an informative abstract could enable them to decide for sure whether or not they wanted to read the entire report. Similar views have been expressed by scientists in other fields.

In some cases, the contents of a report cannot be summarized informatively without making the abstract excessively long. Comprehensive survey reports and proceedings of meetings are obvious candidates for descriptive abstracts. Even for conventional research reports, descriptive abstracts may sometimes be desirable or necessary because of space, cost, time, or other limitations.

---

[4]Gray, Dwight E., "Physics Abstracting," *American Journal of Physics, 18*, 7, 417-424, October 1950.

Writing Your Abstract

The following guidelines concern the style, organization, format, and content of a technical report abstract. Unless specifically qualified, they apply to both the descriptive and informative types.

I. Content

    A. In selecting material for the abstract, keep firmly in mind that a major value of the abstract to the reader (almost the only value for a descriptive abstract) is to help him decide whether or not to read the full report.

    B. Do not repeat the title; material in the abstract should supplement, not duplicate, the title.

    C. Include as a minimum, and usually in this order, concise information on:

        1. The general subject field of your report.

        2. The phases of this general field dealt with in your report.

        3. The approach used in dealing with them, e.g., whether experimental or theoretical.

    D. If the abstract is informative, also summarize, as a minimum:

        1. Newly observed facts.

        2. Degree of accuracy achieved.

        3. Conclusions and recommendations.

    E. In summarizing data and results in an informative abstract, be as quantitative as space and the nature of the material permit; for example, "$-10°F$" requires less space than does "a very low temperature."

II. Style

   A. Use complete sentences, giving particular attention to sentence structure; employ subordination and coordination to the maximum extent compatible with clarity.

   B. Achieve condensation by omitting preliminaries, examples, descriptive details, illustrative incidents, and the like.

   C. Be concise and make every word count, but do not be telegraphic; do not omit a's, an's, and the's.

III. Organization and Format

   A. Remember that a well-written report is by far the most effective first step toward a well-organized abstract; preserve in the abstract, as far as possible, the general organizational scheme of the report.

   B. Make sure that the abstract reflects the same emphasis as the report; otherwise the reader will be misled.

   C. Make sure the abstract is complete in itself, and never make specific reference in it to sections, tables, or illustrations in the report; much of its value is lost if the reader has to read the report in order to understand the abstract.

   D. Hold the length of the abstract to a small percentage of that of the report—not over 5 percent for medium length reports, never more than 10 percent for long documents.

   E. Keep the number of paragraphs in the abstract to an absolute minimum; ordinarily one paragraph will suffice except for an occasional highly informative abstract.

F. Place the abstract, usually on a page by itself, somewhere between the Title Page and the Introduction; the sequence shown in Chapter 1 recommends that it be placed immediately ahead of the Introduction.

As a final step, be sure to have one or more of your colleagues who are knowledgeable in the field of your report read the abstract and comment critically on its form and effectiveness. In professional abstracting circles, a controversy has raged for years over whether the author of a research paper is the best or the least qualified person to abstract it. Those in the pro-author group say he is best qualified because he knows the most about the document and the research it reports. The anti-author people argue that he is much too close to the "trees" to be objective about the "forest" and, therefore, is least fitted for the abstracting task. When you prepare an abstract for a report you have written, you can avoid, or at least minimize, the defects viewed with such alarm by this second group by obtaining constructive criticism from your colleagues.

CHAPTER **8**

Selecting a Title and Preparing the Title Page

That ancient and badly frayed cliché "last but not least" can be appropriately applied to selection of the title and design of the title page of a technical report. If well-phrased and carefully planned, these elements add substantially to the professional stature and effectiveness of the document. If badly done, at best they contribute nothing, and at worst they may mislead and confuse potential readers of the report.

THE TITLE

While you were writing your report, you undoubtedly used, or at least had in mind, some kind of working title. Now you must decide whether it is the best of all possible titles and, therefore, the one you should finally assign your report. If you conclude it isn't, you must devise a better one. In making this assessment, you should consider what the principal function of a report title is, and the criteria that must be met if your title is to perform this function satisfactorily.

Basic Function and Criteria

The principal function of the title of a technical report is to contribute toward accomplishment of the report's overall mission, which is the effective communication of certain technical information. Such, of course, is not the prime purpose of the title in some other forms of literature. For example, "Leaves of Grass," "For Whom the Bell Tolls," "Blood of Strawberries," and the like are excellent titles for the publications they represent, but they do not convey to the reader any appreciable amount of enlightening, substantive information. For a technical report, however, the basic function of the title is to inform the reader, to the extent practical, of the report's subject field and of what distinguishes it from other reports on that same general topic.

The qualification "to the extent practical" marks the point at which trouble invariably arises when one is devising a report title. Ideally, the title should be both brief and sharply definitive of the document's contents—characteristics which, unfortunately, tend to be inherently contradictory. In practice, therefore, the best title is likely to be the optimum compromise between a statement too short to be adequately definitive and one too long and cumbersome to be acceptable as a title.

As an example, consider an actual technical paper

titled "Cold-Starting Tests on Diesel Engines." [1] Conceivably, the author might have simply called his article "Diesel Engines." One would grade this title very high on brevity, but quite low on definitiveness since, for all it indicates to the reader, the paper might be a comprehensive history of Diesel engines, a discussion of Rudolph Diesel and why his first engine blew up, an analysis of the advantages and limitations of such engines in various applications, or the treatment of any of a dozen other aspects of the construction, operation, and use of this type of power plant. On the other hand, if the author had gone all out for defining content without worrying about brevity, he might have entitled the paper "Starting Tests in the Temperature Range -20°F to -70°F Made on Diesel Engines of 1, 3, 4, and 6 Cylinders Using 40- and 60-Centane Fuel"—a statement that approaches being a descriptive abstract. The title he did select is a fair compromise between these extremes.

Other Examples—Bad and Good

Listed below are three groups of titles taken from the title pages of actual technical reports. Those listed first suffer seriously from emphasis on brevity at the expense of giving the reader even a modicum of useful information about the report's content. The second group falls close to the other end of the report title spectrum—that is, they are quite informative but too long and complicated to be satisfactory titles.

[1] *Trans. Soc. Automotive Engineers, 51*, 356-361, October 1943.

Finally, four titles are cited that appear to represent reasonably good compromises of the two basic requirements.

BRIEF BUT NOT SUFFICIENTLY INFORMATIVE

1. Progress Report No. XXIV
2. Built-Up Roofing
3. Technical Gain
4. Review of Work
5. Numerical Control

INFORMATIVE BUT TOO LONG

1. The Influence of Nitrogen Source and Carbohydrate Change by Debudding and Girdling on Bacterial Blight Resistance Caused by the B7 Gene in Cotton

2. On Field Experiments With a Sketch of a Plan for a Wind Wave Generation Experiment to Be Carried Out Off Aruba, N.A. Aboard the R/V Bryely Warfield

3. Aerodynamic Characteristics of Bodies of Revolution at Mach Numbers from 1.50 to 2.86 and Angle of Attack to 180 Degrees

4. Influence of the Turbulent Diffusion Boundary Layer on the Apparent Kinetics of Surface Catalyzed Reactions in External Flow Systems

REASONABLY GOOD COMPROMISES

1. Wear Characteristics of Metal-Polymer Friction Pairs
2. Thermal Conductivity of Saturated Leda Clay
3. Effect of Energy Changes on Solar Cosmic Rays

4. Structure of Normal Shock Waves in a Binary Gas Mixture

A title that is either unduly cryptic or excessively long and involved usually indicates that the author was both a bit indolent and somewhat lacking in understanding of a report title's real function. By devoting a little careful thought and effort to the task, you will be able to devise a title for your report that combines reasonable brevity and conciseness with enough definitive description of content to have real value for your readers.

THE TITLE PAGE

Fellow engineers who read your report, administrators who scan it to decide whether they want to read it, and librarians who have to catalog and circulate it—all of these groups deserve to find readily available in the report, a concise statement of what one may call its "vital statistics." In good technical reports, this information, or most of it, appears at the very beginning on a sheet called the Title Page. Discussed below are the kinds of data this page should present, how this information should be arranged, and the relationship of the title page to a supplementary Bibliographic Control Sheet, which is sometimes included, and to the document's cover.

Title Page Information

Here, one may paraphrase the traditional requirements for the lead paragraph of a news story and say that the title page of a technical report *always* should answer at least the questions of What?, Where?, When?, By whom?, and To whom? More specifically, the title page of your report should state, as an absolute minimum:

1. Complete title of the document. (Recurring titles, as in progress report series, should be consistent.)
2. Name, position, and organizational affiliations of the author or authors. (If personal writers are not to be named, the originating organization or corporate author should be identified.)
3. Name of the person or organization, or at least the organization, to whom or to which the report is being submitted.
4. Date of issuance of the report.
5. Any identifying numbers that have been assigned by the time the report is printed.
6. Security classification (if applicable).

In addition to the above "must" items, several other kinds of information are often appropriate for the title page or the bibliographic control sheet that follows it (see below). These principally include:

1. Type of report (preliminary, interim, progress, final, other)
2. Approvals by administrative officials
3. Period covered by the work
4. Limitations on distribution
5. Routing schedule
6. Contract and project numbers

7. Names of cooperating, sponsoring, or co-sponsoring agencies

(Note that these items are *in addition* to the group listed above, and should be included only if applicable in the particular case.)

In developing these lists of title-page elements, we consulted with Library of Congress catalogers who had cataloged many thousands of technical reports. Obviously, one would not argue that cataloging convenience should be the overriding consideration in planning a report title page. Nevertheless, it seems fair to assume that a title page presented in the form best suited to the requirements of librarians, who have to analyze reports descriptively, will also be more or less ideally organized for the use of individuals who read, scan, store, and route reports. For your report, therefore, include in your title-page data, in addition to the six essential elements, any of those in the second group that are applicable to your reporting situation. The problem of what to do if there are too many of these for one page is discussed below.

Format of the Title Page

Within a wide range of patterns, there is no absolutely right or wrong way to arrange the above kinds of information on your title page. Many agencies have standardized formats that must be used with all the reports they issue. If this is the case in your organization, you have no problem. If, however, you have to design your own title page, probably as good a rule as any to follow is the precept laid down by Sherman:

"The material on the title page should be pleasingly arranged, with emphasis on the upper half of the page." [2]

Bibliographic Control Sheet

Sometimes it is desirable to include, immediately following the title page, a supplementary page known as the bibliographic control sheet. The need for this sheet usually arises under one, or both, of the following conditions:

1. The agency's standardized title-page format does not provide for all of the descriptive data the author wishes to present. *OR*
2. The total amount of such information is too great for one page.

If you are in this situation with regard to your report, you must decide how to divide these elements between the title page and the bibliographic control sheet. Any standardized title-page format that you may have to follow almost certainly will provide for at least the five or six elements described above as absolutely essential; if it doesn't, it should. If you design your own title page, make certain they appear on it. If approvals by administrative officials are to be shown, these too are usually stated on the title page. Items in the second group that clearly belong on the bibliographic control sheet are those that relate specifically to the bibliographic control and distribution

[2]Sherman, Theodore A., *Modern Technical Writing*, Prentice-Hall, Englewood Cliffs, N.J., 1966, p. 185.

of the document—for example, limitations on distribution and routing schedules. Divide the others among the two sheets in whatever manner you think will serve your readers best.

Relation of the Title Page to the Cover

Most technical reports that are more than a few pages in length have special covers. These serve to protect the contents and, it is hoped, to add to the document's professional appearance and the prestige of the issuing agency. If your report is to have such a cover, it will of course carry some of the same information as that slated above for the title page. Most report covers give, *as a minimum*, the title, author (personal and/or corporate), date, identifying numbers, and security classification (if any). Note, however, that the appearance of this information on the cover in no way affects what should be on the title page and bibliographic control sheet described above. In short, let the cover be whatever established agency procedures decree, but make certain that the title page and bibliographic control sheet (if included) present your report's vital statistics clearly, concisely, and accurately, for the benefit of those who will read, catalog, route, and store your report.

CHAPTER 9

Review of the Characteristic of a Good Technical Report

The emphasis in the first eight chapters of this book has been on the mechanics and ground rules of putting a technical report together, although quality factors have not been entirely ignored. Chapter 9 reverses the emphasis and concentrates on certain quality characteristics that are important to the achievement of a high degree of technical report effectiveness. Specifically, the attributes considered are: completeness, clarity, conciseness, veracity, restraint, and general appearance.

Chapter 9 is called a review because most of these attributes of a good report have been mentioned earlier—principally in Chapters 1 and 2. Emphasis on them through amplification of the previous brief references is appropriate at the close of our discussion of the overall technical report-writing problem.

COMPLETENESS

Two aspects of technical report completeness are considered in this section. One relates to the scope of the subject matter presented in the document, the other to its internal continuity and tidiness.

Scope of Subject Matter

As far as is practical, a technical report should stand on its own feet. That is to say, ideally every reader in the group to whom the report is directed should be able to follow its line of reasoning without consulting other sources. As we saw earlier, however, it is difficult to define precisely how much background information reader groups possess. Even within the feasible framework of approximate definition, a given technical report may have to serve several different reader groups. To complicate the matter still further, the amount of background information that should be supplied is, to some degree, a function of the particular content. For example, suppose that in the body of a report the writer has had occasion to apply the following ancient relationship:

> *The square of the hypotenuse of a right-angled triangle equals the sum of the squares of the other two sides.*

This celebrated theorem is much too well known for it to be necessary for the author to include Pythagoras's famous derivation in his report. On the other hand, equations or principles that are less familiar often should be supported with derivations or other background data, either at the point in the report where the application occurs or in an appendix. Thus, once again, the writer must use his own good judgment and common sense. He should ask himself (and thoughtfully answer) the question, "Just what, and how much, must I include by way of background, to give those for whom I am writing this report substantially the complete story?" As we have said before, in case of doubt, erring a bit on the side of presenting too much information is preferable to taking a chance on including too little.

Internal Continuity

All too frequently, when the reader of a technical report is riding smoothly along with the author's train of thought, he suddenly encounters the mental equivalent of a missing section of track or an unbridged canyon. True, the report right-of-way usually picks up again on the other side of the chasm, but the reader may have been given no structural material to use in bridging it. Or, without warning, he may find that the writer has neatly switched him off onto a blind siding in the argument, without making provision for getting him back on the main line. Discontinuities of these kinds are most likely to occur when the report is written solely by the individual responsible for the work being reported. The document may read logically and appear complete to the author because, as he goes along, he subconsciously supplies the missing bridges and rails from his first-hand knowledge of what he did and why and how he did it.

The lack of continuity results primarily from his un-
conscious failure to put himself fully in the position
of the reader.

CLARITY AND CONCISENESS

A paragraph, or series of paragraphs, can be clear
without being highly concise; in general, however, the
more concise the passage, the easier it is to comprehend. Conversely, expository writing can be concise
and still be difficult to understand, but if it were not
concise, the meaning undoubtedly would be even
more abstruse. In short, although the terms clarity
and conciseness are not synonymous, they are closely
enough related—at least in technical report writing—
for us to consider them together in this chapter.
Among the principal enemies of clarity and conciseness are unnecessary material, obscure phrasing,
overly complex sentences, inexact usage, and illogical
organization. We shall consider and illustrate these
somewhat overlapping factors in three groups.

Unnecessary Material

Coleridge once said that the art of writing consists
largely in knowing what to leave in the inkwell. Technical report content in the fugitive-from-an-inkwell
category is principally of two general kinds: information that the reader doesn't really need but which the
author is tempted to give anyway, and non-working
wordage.

It is stated in Chapter 1 that a technical report
should contain all the data and discussion necessary

for the reader to get the message and, ideally, nothing more. To achieve this goal, even approximately, the writer must, first, distinguish between essential and non-essential content and, second, force himself to omit that which is not pertinent. The former ordinarily presents no great problem, providing the author gives the matter careful thought; the latter, however, may prove quite difficult, especially if he is deeply interested in some particular aspect of the report's subject matter.

In this connection, Sherman[1] cites the case of a report on a study made to determine whether or not the grasshoppers in a certain farming region should be poisoned and, if so, how this deed might best be accomplished. In his report, the author dealt with crop damage by grasshoppers, cost of poisoning, relative efficiency of various poisoning techniques, and other appropriate aspects of the problem. Then, being personally greatly interested in, and an authority on, grasshoppers, he proceeded to include an exhaustive, almost totally irrelevant, discussion of the life cycle and miscellaneous habits of these insects. It is always difficult to resist an opportunity to ride one's own hobby. When a technical report writer succumbs to this temptation, he always reduces the effectiveness of his writing.

The type of unnecessary material just discussed lowers the impact of a technical report because it dilutes legitimate content with irrelevant information. The other category of unnecessary material consists

---

[1] Sherman, Theodore A., *Modern Technical Writing*, Prentice-Hall, Englewood Cliffs, N.J., 1955, 1st ed., p. 129.

of what are sometimes called "idle words"—that is, words and phrases that make no useful contribution and, therefore, simply clutter up the document and make it more difficult both to read and to understand. The following pairs of examples illustrate several of the most common kinds of idle-word phenomena. In each case, the passage is given, first, as it actually appeared in a document and, second, as it might have been phrased with the non-essential words deleted.

As Written	*As Might Have Been Written*
We have made enough progress in so far as that we at least recognize etc.	We have made enough progress to recognize etc. *OR* We now recognize etc.
I must make it plain, of course, that further tests are necessary for the simple reason that etc.	Further tests are needed because etc.
This is to inform you that the report should be sent directly to this office as promptly as possible.	The report should be sent to this office promptly. *OR* Send the report to this office promptly.
Attached hereto *OR* Enclosed herewith etc.	Attached *OR* Enclosed etc.
It will be noted that all memoranda emanating from the New York office etc.	All memoranda from the New York office etc.
There are two questions which must be answered.	Two questions must be answered.
Your attention is directed to Chapter 8 which says etc.	Chapter 8 says etc.

As Written	*As Might Have Been Written*
Additional supporting data, information, comments, and supporting documentation may be included beneath the writing referred to above, as deemed necessary.	Supporting data may be appended.

What causes technical report authors, and others, to use "idle words"? A variety of reasons, including one or more of the following, are responsible: the writer's mistaken notion that the wordier a passage is, the more impressive it will read; his desire, often not consciously recognized, to let responsibility for the statement remain a bit fuzzy; a certain degree of doubt in his mind about precisely what he wants to say; and a combination of laziness and carelessness in deciding how to phrase his story.

Obscure Phrasing, Inexact Usage, Overly Complex Sentences

These shortcomings are not mutually exclusive, and any passage that exhibits one of them is likely to suffer from the other two as well. The basic principle involved is stated in Chapter 1: "Effective performance of the technical report's principal mission makes it imperative that the writing be clear, concise, precise, and phrased in unambiguous, functional English." Almost any reader's reaction to the following quotations from actual documents will demonstrate the validity of this thesis:

1. From a technical report:

This is due to the fact that when there is a steep temperature gradient in the steel the higher conductivity of the gold causes it to have a much lower gradient for the same total rate of heat flow so that a layer of gold will have its surfaces more nearly at the same temperature than would exist at the same points if the intervening material were steel.

2. From a university president's justification for a new school of communication:

The ability to utilize the techniques of communication provided by the technology of our age for the clear and rapid dissemination of information and the ability to draw upon the scholarship and arts of our institutions of higher education to reduce the incidence of semantic ambiguity and demogogic device require the existence of a skilled and educated profession of communications.

3. From an encyclopedia article on "The Ascent and Descent of Air":

While the lapse rate remains on the average less than the dry adiabatic, but slightly greater than the saturated adiabatic, we can readily conceive of any isolated mass of air which has become saturated and at a slightly higher temperature than its environment being able to rise through its environment, since in the circumstances postulated its temperature would be at each successive level higher than that of its immediate environment. The converse process of the descent of air, however, is not readily understandable.

(This mish-mash was revised in a subsequent edition.)

4. From testimony before a Congressional appropriations committee:

We try to maintain a balance between the hard, as it is called colloquially, which means, in many instances, subterranean construction, and the effectiveness of our deterrent forces on the other hand which apparently, under present concepts, is regarded as the best instrumentality as opposed to a hidden defense posture.

One exasperated member of the Congressional committee to whom this last statement was made said to the witness, "You are wonderful. I haven't the faintest idea what you said. You just spew out words. What in the world do you mean?" Such might well have been the response of hearers and readers of all of the above examples of cloudy, obscure, non-English.

Jordan Brotman has written, "In good technical papers the writing is completely transparent; nothing stands in the way between the reader and your [the writer's] ideas. Any uncertainty in your way of putting things makes the reader conscious of the medium you are using. He doesn't have to be a grammarian. He has read enough to sense the difference between good and clumsy writing. If your writing is clear he has confidence in you." [2] Meandering sentences, wheels-within-wheels construction, imprecise usage, circuitous phraseology, irrelevant data, and non-pertinent discussion all tend to confuse and frustrate the reader and to destroy his confidence in the writer. In short, to adapt the plea which, in a different context, the queen in *King Richard II* makes

[2] Lawrence Radiation Laboratory (Livermore), Technical Information Division Bulletin, No. 1, 21 August 1961, 3p.

to the Duke of York, "Uncle, for God's sake, speak comfortable words." [3]

Logical Organization

Illogical arrangement of material can wreck the effectiveness of a document quite as completely as can any of the shortcomings discussed above. If the reader is to obtain satisfaction in reading and give his full attention to content and argument as he makes his way through a technical report, point A must lead logically to point B, B to C, C to D, and so on. If the author jumps from X to Z, the reader is not likely to comprehend the message, and it is even less probable that he will accept the conclusion. Then if he comes across the missing link Y at some later point in the report, he may not recognize it as such, or, if he does, his reaction is sure to be a frustrated "Now they tell me!"

As is emphasized in Chapter 2, a report author should begin to worry about the logical organization of the material he plans to present when he is first developing his working outline. If it is well organized when he starts to write, and if he maintains the tidiness of his argument as he goes along, the sequence of presentation in his completed report will probably make sense to his readers.

VERACITY AND RESTRAINT

Although the relationship between veracity and restraint is of a somewhat different kind than that be-

[3] Shakespeare, William, *King Richard II,* Act II, Scene 2, Line 76.

tween clarity and conciseness, it is sufficiently close to warrant our discussing them together.

Veracity

As the term is used here, veracity implies much more than mere accuracy in mathematical calculation and in transcribing data from the writer's notebook to his report manuscript. It means, for example, drawing only conclusions that are warranted by the facts and figures presented in the report, and making only recommendations that are justified by the specific conclusions drawn in the report. It includes taking care to avoid the possibility of erroneous inferences by the reader because of the omission of pertinent data. And, closely related to exact usage discussed earlier it means stating facts in quantitative rather than qualitative terms wherever possible—for example, one teaspoon of salt, not a pinch of salt; or a 75 percent increase, not a large increase. Veracity can be defined as "conformity to truth." A technical report possesses this characteristic to the degree that what it says and what it omits combine to give the reader an accurate understanding of the writer's intended message.

Restraint

We have linked this factor to veracity in this discussion because lack of restraint by the writer is a major cause of the failure of technical reports to be completely truthful. The report author who exercises restraint seldom, if ever, uses such adjectives as revolutionary, sensational, enormous, fabulous, and the like. Instead, as far as possible, he presents quantitative values, allowing the reader to make his own judgment as to whether they are indeed revolutionary, sensational, and so forth. If he has to employ qualitative terms, he favors words like substantial, appreciable, and their similarly conservative cousins, over the more flamboyant and superlative members of the family. He avoids the kind of glowing, but not always completely accurate, language that, for example, is not entirely unknown in advertising claims for weed killers, soap powders, cosmetics, and headache remedies. Since restraint is closely related to objectivity, a technical report writer is in little danger of erring seriously in this regard if all of his conclusions and recommendations are intimately, carefully, and objectively correlated with the actual data and results presented in his report.

FINAL CHECKS ON THE MANUSCRIPT

Your report-writing task is now almost finished. Your goal has been to tell a complete story in clear, concise, unambiguous, functional English. Two final exercises will help ensure that you have achieved this objective.

The first is simply to thoughtfully read the manuscript yourself, from beginning to end, *in one sitting.* While writing, you were working on, and thinking about, the material one paragraph, subsection, section, or other fragment at a time. An uninterrupted read-through of the report as a whole is certain to bring to light gaps in logic, awkward phraseology, inconsistent statements, and other defects that previously escaped your notice. It has been this writer's experience that doing the reading aloud improves the effectiveness of this exercise.

The other final check requires you to obtain help from your friends. It probably is impossible for any scientist or engineer, reading a report he wrote on an experiment he personally conducted, to completely separate what he sees on the printed page from what he knows because he was the experimenter. The remedy is for him to have one or more (preferably more) knowledgeable colleagues read and criticize his manuscript. For the best results, these colleague critics should be competent in the subject field of the report, but should not themselves have participated directly in the particular work the document describes.

GENERAL APPEARANCE

To be fully effective, a technical report should be at least reasonably neat and attractive in appearance. A report can contain all of the proper ingredients, can be well written with its theme logically developed, and can still fall short of getting its message across, if externally and internally it looks messy and unattractive. At the other extreme, ornate covers, multicolor printing, complex overlays, and the like, are sometimes used to conceal lack of substance or to satisfy the issuing organization's exhibitionistic urges, and for essentially no other reason. But an expensive, highly elaborate production job is not necessary. Between these extremes is the report that is good in terms of what it contains and is legibly printed, with the tables well presented, the illustrations tidily drawn, and the cover simple, neat, and attractive. Such a technical report can approach the maximum in effectiveness. The writer with a significant story to tell who issues his report in a format that is untidy, forbidding looking, and difficult to read unnecessarily handicaps himself and jeopardizes his chances of communicating effectively with his readers.

CHAPTER 10

Capsule Summary of Chapters 1-9

I. Nature and anatomy of a technical report.

   A. An effective technical report (pp. 1-7),

   1. Communicates technical information accurately, precisely, and readably.

   2. Is directed primarily toward a specific group of readers who have use for it, have asked for it, and/or have a right to expect it.

   3. Is organized in the form best suited to the readers' needs.

   B. A representative technical report might contain the following sections, *arranged in this order in the completed document*: (See the rest of this Summary for the recommended sequence of preparation.)

   1. Title—Should be brief and as definitive as possible of the report's contents. (pp. 69-73)

   2. Title Page—Should give, as a minimum: Title, name and position of author, person or

organization to whom report is directed, date of issuance, identifying number(s), security classification (if any). (pp. 73-76)

3. Summary of Conclusions and Recommendations—May or may not be included depending upon the nature of the report. (pp. 45-46)

4. Table of Contents—Should be included in any report more than a few pages in length. (pp. 57-58)

5. Lists of Tables and Illustrations—Should be included if there are more than one or two of each. (p. 58)

6. Abstract—Should be included in any report more than two or three pages in length; may be descriptive, informative, or something in between. (pp. 59-68)

7. Introduction—Should introduce the reader to the report's subject matter, and orient him for an orderly, satisfying perusal of the rest of the document. (pp. 23-30)

8. Body of Report—Never appears as an actual report section heading; term is used here to encompass sections on what was done, how it was done, and what was learned. (pp. 31-39)

9. Conclusions—May or may not be included depending upon the nature of the report; if present, should stem directly from data and results in the body of the report. (pp. 40-46)

10. Recommendations—May or may not be included depending upon the nature of the report; if present, should stem directly from the Conclusions. (pp. 40-46)

11. Appendixes—May or may not be

needed; suitable for supplementary information not essential to report's main thesis, but probably useful to the reader as background material. (pp. 47-50)

    12. Bibliography or List of References— May include documents specifically referred to in the text, supplementary reading material, or both. (pp. 50-54)

    II. Getting ready to write a technical report— that is, before you start to write,

    A. Analyze the overall problem, which means giving careful thought to determining,

    1. Exactly what elements of information you wish to present. (pp. 11-14, 32-37)

    2. Exactly for whom you are writing the report. (pp. 10-13)

    3. Precisely how much history, background, theory, and the like you wish to include. (pp. 10-11, 33)

    4. The most logical sequence in which to present the material. (pp. 15, 37-39)

    B. Gather together the information you wish to present. (pp. 11-14, 32-37)

    C. Make a detailed working outline. (pp. 14-21)

    D. Plan your tables and illustrations, if any. (p. 58)

    E. Give the ideas you have generated in steps A through D some time to ferment in your mind—that is, sit and think for a while. (pp. 21-22)

    F. Do not be surprised if this pre-writing phase takes as long as, or longer than, the actual writing—it probably should.

III. Writing the introduction. (NOTE: From here on, the sections are considered in the sequence in which it is recommended they be written.)

    A. Be sure your introduction,

        1. Makes clear to the reader the precise subject(s) the report considers. (p. 25)

        2. Acquaints him with the particular purpose of your report. (pp. 25-26)

        3. Summarizes the plan according to which you are presenting the material—that is, includes the "road map." (pp. 26-28)

    B. Decide whether or not you should include any historical or other background information in the introduction; do so only if you believe the report needs it. (p. 29)

    C. Take particular care with the initial sentence of your introduction—it can go far toward setting the pattern of reader acceptance. (p. 28)

IV. Writing the body of the report.

    A. Now that you have written the introduction, re-check your working outline to make certain,

        1. That the main-body section you tentatively selected will inform your readers of what you did, how you did it, and what you learned. (pp. 14-21)

        2. That they are arranged in the sequence which, in your opinion, will make the best sense *from the reader's standpoint.* (pp. 14-21)

    B. Re-check your earlier decision regarding tables and illustrations. (p. 58)

    C. Write the main-body sections. (pp. 31-39)

V. Writing the conclusions and recommendations.

A. Regarding your conclusions (pp. 40-46),

1. Make certain they follow logically
from the results and argument presented in the main
body.

2. Make certain they are consistent with
what your introduction promised regarding the report's contents.

3. Decide whether or not you wish to
summarize the conclusions at the beginning of the
report.

B. Regarding recommendations (pp. 40-46),

1. Make certain they follow directly
from the conclusions.

2. Same as V. A-2 and A-3 above.

VI. Preparing the appendixes, references, and
contents.

A. Regarding appendixes (pp. 47-50),

1. Principal criterion that material in
appendixes should meet is: Not essential to the main
argument of the report, but probably of supplemental
value to the reader.

2. Specifically, typical appendix-type
possibilities include: Derivations of equations, historical summaries, expansion of tables and data lists, supporting computations, description of processes that
failed, and the like.

B. Regarding references and footnotes (pp.
50-56),

1. Decide whether or not to distinguish
between a list of references (items to which specific

reference is made in the text of the report) and a strict bibliography (of non-referenced, supplementary reading).

    2. Select one of the accepted forms of citation and use it for all references.

    3. Place footnotes at the bottoms of the pages where the identifying asterisks or superscript numbers occur; for those that are documents, use the same form of citation as for references.

    C. Regarding the table of contents (pp. 57-58),

    1. Include such a table if the report is longer than two or three pages.

    2. Use topical headings from the final form of your detailed outline; usually, one or two "decks" of heading suffice in the table of contents.

    3. Make certain that every item in the table of contents appears as a section or subsection heading in the report; the reverse need not be true.

    D. Regarding lists of tables and illustrations (p. 58),

    1. Include such a list, or lists, if the report contains more than one or two tables or figures.

    2. For each figure or table, give the number, title, and page reference.

    VII. Writing the abstract.

    A. Decide precisely where you want the abstract of your report to fall along the spectrum from purely descriptive to highly informative. (pp. 59-60, 64)

    B. To the extent that it is to be descriptive, make certain it will enable the reader to decide whether or not he is interested in reading the report. (pp. 60-61)

C. To the extent that it is to be informative, make certain it reflects the same emphasis as the report, and gives the reader the principal results, conclusions, and/or recommendations. (p. 61)

D. Write the abstract to stand on its own feet. (p. 67)

E. Be as economical of words as you can without being telegraphic, and make every word count; the better the working outline, the easier it is to write the abstract. (p. 67)

VIII. Selecting a title and preparing the title page.
    A. Regarding the title (pp. 69-73),
      1. Select the optimum compromise between "informative but too long" and "short but too general."
      2. Make the title reflect the basic emphasis of the report.
    B. Regarding the title page (pp. 73-77),
      1. Make certain it carries *as a minimum*: Complete title, name of author and his position, name of person or organization to whom the report is being submitted, date of issuance, identifying number(s) (if any), and security classification (if applicable).
      2. Include other items as circumstances require; if there are too many for one page, add a supplementary bibliographic control sheet.

IX. Characteristics of a good technical report.
    A. Completeness—Should possess internal

continuity, that is, it should tell a complete story. (pp. 79-81)

B. Clarity and conciseness—Should, as far as possible, contain everything necessary for the reader to get the message and nothing more; should be logically organized and written in clear, concise, precise, unambiguous, functional English. (pp. 81-87)

C. Veracity and restraint—Should conform to the truth, with regard both to the accuracy of data and results, and to the relationship of the conclusions and recommendations to these results. (pp. 87-89)

D. Attractive appearance—Should be highly legible, and be presented in a well-planned, tidy, attractive format. (pp. 90-91)

Appendixes

Selected Bibliography:
Style and Usage

As their headings indicate, neither this bibliography nor the one that follows it is meant to be comprehensive. The volumes listed below are representative of authoritative publications on English style and usage, and include a number of books this author has found particularly helpful. In addition to their value as references, many of them are actually fun to read. Two further points worth keeping in mind are:

- A sizable fraction of the books cited in any bibliography themselves contain bibliographies or lists of references, with many of the entries in these compilations including, in turn, further such lists, and so on.
- Many leading professional societies publish specialized guides to style and usage for publications in their respective subject fields.

1. Bernstein, Theodore M. *The Careful Writer: A Modern Guide to English Usage.* Atheneum, New York, 1965. 487 p.
2. Copperud, Roy H. *Words on Paper: A Manual of Prose Style.* Hawthorn, New York, 1960. 286 p.
3. COSATI. *Guidelines to Format Standards for Scientific and Technical Reports Prepared By or For the Federal Government.* Clearinghouse for Federal Scientific and Technical Information, Springfield, Va., 1968. 22 p.
4. Evans, Bergen. *Comfortable Words.* Random House, New York, 1962. 379 p.
5. Evans, Bergen and Cornelia Evans. *A Dictionary of Contemporary American Usage.* Random House, New York, 1957. 567 p.
6. Flesch, Rudolph. *The ABC of Style.* Harper and Row, New York, 1964. 303 p.
7. Fowler, Henry Watson. *A Dictionary of Modern English Usage.* Rev. by Sir Ernest Gowers. Oxford, New York, 2nd ed., 1965. 725 p. (See also No. 13.)
8. Gunning, Robert. *The Technique of Clear Writing.* Rev. ed. McGraw-Hill, New York, 1968. 329 p.
9. Lambuth, David, et al. *The Golden Book on Writing.* Viking Press, New York, 1964. 81 p. (Paperback)
10. Linton, Calvin D. *Effective Revenue Writing I and II.* U.S. Government Printing Office, Superintendent of Documents, Washington. Pt. I, 1961, 261 p.; Pt. II, 1962, 198 p. (Paperbacks)
11. *A Manual of Style.* University of Chicago Press, Chicago, 12th ed., 1969. 546 p.
12. National Academy of Sciences. *A Guide for Preparing Manuscripts.* Washington, 1970. 60 p.
13. Nicholson, Margaret. *A Dictionary of American-English Usage.* Oxford, London, 1957. 671 p. (Based on Fowler's *A Dictionary of Modern English Usage*—No. 7 above)

14. Nicholson, Margaret. *A Practical Style Guide for Authors and Editors*. Holt, Rinehart, and Winston, New York, 1967. 143 p.
15. O'Hayre, John. *Gobbledygook Has Gotta Go*. U.S. Government Printing Office, Superintendent of Documents, Washington, 1966. 113 p.
16. Roget, Peter Mark. *International Thesaurus*. Crowell, New York, 1962. 1258 p.
17. Roget, Peter Mark. Ed. by Norman Lewis. *The New Roget's Thesaurus*. Putnam, New York, 1964. 552 p.
18. Skillin, Marjorie E., Robert M. Gay, et al. *Words Into Type*. Appleton-Century-Crofts, New York, 1964. 586 p.
19. Strunk, William, Jr., and E. B. White. *The Elements of Style*. Macmillan, New York, 1959. 71 p. (Hard cover and paperback)
20. Turabian, Kate L. *A Manual for Writers of Term Papers, Theses, and Dissertations*. University of Chicago Press, Chicago, 1967. 164 p.
21. Turner, Rufus P. *Technical Writer's and Editor's Stylebook*. Bobbs-Merrill, New York, 1964. 208 p.
22. U.S. Government Printing Office. *Style Manual*. Rev. ed. Superintendent of Documents, Washington, 1967. 512 p.
23. Wallace, John D. and J. Brewster Holding. *Guide to Writing and Style*. Battelle Memorial Institute, Columbus, Ohio, 1966. 115 p. (Paperback)

Selected Bibliography: Technical Writing

As in Appendix A, the objective of this Appendix is to list a limited, representative group of authoritative publications particularly useful to technical report authors. Again, it should be remembered that books listed in bibliographies contain short bibliographies that cite still other volumes with bibliographies or lists of references. Thus, even a limited compilation can lead one to a bibliography that is quite comprehensive.

1. Cooper, Bruce M. *Writing Technical Reports*. Penguin, Baltimore, 1964. 190 p. (Paperback)
2. Crouch, W. G. and R. L. Zetler. *Guide to Technical Writing*. Ronald Press, New York, 3rd ed., 1964. 447 p.
3. Douglas, Paul. *Communication Through Reports*. Prentice-Hall, Englewood Cliffs, N.J., 1957. 410 p.
4. Gilman, William. *The Language of Science: A Guide to Effective Writing*. Harcourt, Brace, and World, New York, 1961. 248 p.

5. Graves, Harold F. and Lynn S. S. Hoffman. *Report Writing*. Prentice-Hall, Englewood Cliffs, N.J., 4th ed., 1965. 286 p.

6. Hicks, Tyler G. *Writing for Engineering and Science*. McGraw-Hill, New York, 1961. 450 p.

7. Houp, Kenneth W. and Thomas E. Pearsall. *Reporting Technical Information*. Glencoe Press, Beverly Hills, Calif., 1968. 382 p. (Paperback)

8. Jones, W. Paul. *Writing Scientific Papers and Reports*. Wm. C. Brown, Dubuque, Iowa, 5th ed., 1965. 288 p. (Paperback)

9. Kapp, Reginald O. *The Presentation of Technical Information*. Macmillan, New York, 1957. 147 p.

10. King, Lester S. and Charles G. Roland. *Scientific Writing*. American Medical Association, Chicago, 1968. 132 p. (Paperback)

11. Menzel, Donald H., Howard Mumford Jones, and Lyle G. Boyd. *Writing a Technical Paper*. McGraw-Hill, New York, 1961. 132 p. (Paperback)

12. Miller, Walter J. and Leo A. Saidla, eds. *Engineers as Writers*. D. Van Nostrand, New York, 1953. 340 p.

13. Mitchell, John. *A First Course in Technical Writing*. Chapman and Hall, London, 1967. 180 p.

14. Mitchell, John. *Handbook of Technical Communication*. Wadsworth, Belmont, Calif., 1962. 321 p.

15. Nelson, J. Raleigh. *Writing the Technical Report*. McGraw-Hill, New York, 1952. 355 p.

16. Peterson, Martin S. *Scientific Thinking and Scientific Writing*. Reinhold, New York, 1961. 215 p.

17. Rathbone, Robert R. *Communicating Technical Information*. Addison-Wesley, Reading, Mass., 1966. 104 p. (Paperback)

18. Rathbone, Robert R. and James B. Stone. *A Writer's Guide for Engineers and Scientists*. Prentice-Hall, Englewood Cliffs, N.J., 1962. 348 p.

19. Robertson, W. S. and W. D. Siddle. *Technical Writing and Presentation.* Pergamon Press, New York, 1966. 118 p.
20. Schultz, Howard and Robert G. Webster. *Technical Report Writing: A Manual and Sourcebook.* D. McKay Co., New York, 1962. 359 p.
21. Sherman, Theodore A. *Modern Technical Writing.* Prentice-Hall, Englewood Cliffs, N.J., 2nd ed., 1966. 418 p.
22. Smith, Richard W. *Technical Writing: A Guide to Manuals, Reports, Proposals, Articles, Etc.* Barnes and Noble, New York, 1963. 181 p. (Paperback)
23. Society of Technical Writers and Publishers. *An Annotated Bibliography on Technical Writing, Editing, Graphics, and Publishing.* Washington, 1965. 200 p.
24. Turner, Rufus P. *Technical Report Writing.* Holt, Rinehart, and Winston, New York, 1965. 210 p. (Paperback)
25. Ulman, Joseph N. and Jay R. Gould. *Technical Reporting.* Henry Holt, New York, 1959. 382 p.
26. Walton, Thomas F. *Technical Manual Writing and Administration.* McGraw-Hill, New York, 1968. 383 p.
27. Ward, Ritchie R. *Practical Technical Writing.* Knopf, New York, 1968. 270 p.
28. Weil, Benjamin H. *The Technical Report.* Reinhold, New York, 1954. 485 p.
29. Weisman, Herman M. *Basic Technical Writing.* Chas. E. Merrill, Columbus, Ohio, 1962. 512 p.
30. Weisman, Herman M. *Technical Report Writing.* Chas. E. Merrill, Columbus, Ohio, 1966. 200 p. (Paperback)

Manuscript Typing Instructions[1]

The following guidelines are intended primarily for use by the typist who must prepare a technical report manuscript in final form for submission to a printer or publisher. They are good rules to follow regardless of whether the manuscript is to be set up in type and formally printed, or to be reproduced for interoffice use only.

GENERAL

Type the manuscript on one side only of 8-1/2"x 11" (for government, 8" x 10-1/2") opaque, white paper. Double space all typewritten material, including footnotes, quotations, and references. Leave additional space around mathematical equations.

---

[1] Adapted from *A Guide for Preparing Manuscripts,* National Academy of Sciences, Washington, D.C., 1970, pp. 47-49.

Allow liberal margins—about 1-1/2" at left and top, 1-1/4" at right and bottom. As far as possible avoid breaking a word at the end of a line, even if this requires letting the line run a little too long or too short. Ordinarily begin each chapter on a new page.

Unless specifically instructed otherwise, submit the original and one copy to the printer or publisher. Always retain at least one copy in the originating office or unit.

CORRECTIONS

For minor corrections, cross out the error and type, or write legibly, the correct version above it. If not too much material must be changed, the use of correction tape or fluid (e.g., Wite-Out or Snopake) is acceptable. Do not simply type a correct letter over the incorrect one; the result usually is not legible. If a substantial fraction of a page requires correcting, retype the page.

INSERTS

If an insert is no more than a word or two in length, type it or write it legibly between the lines at the appropriate place in the text. Type longer inserts on separate sheets of paper, key them to the points where the additions are to be made, and place each insert sheet immediately after the manuscript page to which the insert is to be added. Avoid extensive hand-written additions and revisions. Do not write in the margins or on the backs of pages. Do not staple or paste bits of paper over manuscript sheets. Do not use tape that covers any part of the text.

QUOTATIONS AND FOOTNOTES

Double space all quotations. If a quotation is short, run it in with the text and enclose it in quotation marks. Place a long quotation in a separate paragraph, indented to set it off from the text; in this case, quotation marks are not necessary.

Double space all footnotes. Either group them at the bottom of the pages on which they are referenced, or insert each one immediately following the superscript or other index mark that refers to it. In the former case, draw a short horizontal line on the page to separate the text material from the one or more footnotes. If footnotes are placed individually within the text, type or draw page-wide, solid lines immediately above and below each footnote.

Index